Defense Planning in a Decade of Change

Lessons from the Base Force, Bottom-Up Review, and Quadrennial Defense Review

Eric V. Larson, David T. Orletsky, Kristin Leuschner

WITHDRAWN BY THE
UNIVERSITY OF MICHIGAN

Prepared for the United States Air Force
Approved for public release; distribution unlimited

Project AIR FORCE **RAND**

The research reported here was sponsored by the United States Air Force under Contract F49642-01-C-0003. Further information may be obtained from the Stategic Planning Division, Directorate of Plans, Hq USAF.

Library of Congress Cataloging-in-Publication Data

Larson, Eric V. (Eric Victor), 1957-
  Defense planning in a decade of change : lessons from the base force, bottom-up review, and quadrennial defense review / Eric V. Larson, David Orletsky, Kristin Leuschner.
    p. cm.
  MR-1387
  Includes bibliographical references.
  ISBN 0-8330-3024-8
  1. Military planning—United States—History—20th century. I. Orletsky, David T., 1963– II. Leuschner, Kristin. III. Title.

U153 .L37 2001
355'.033573—dc21

2001031975

Top left cover photo courtesy of LockheedMartin. Reprinted by permission.

RAND is a nonprofit institution that helps improve policy and decisionmaking through research and analysis. RAND® is a registered trademark. RAND's publications do not necessarily reflect the opinions or policies of its research sponsors.

© Copyright 2001 RAND

All rights reserved. No part of this book may be reproduced in any form by any electronic or mechanical means (including photocopying, recording, or information storage and retrieval) without permission in writing from RAND.

Published 2001 by RAND
1700 Main Street, P.O. Box 2138, Santa Monica, CA 90407-2138
1200 South Hayes Street, Arlington, VA 22202-5050
201 North Craig Street, Suite 102, Pittsburgh, PA 15213
RAND URL: http://www.rand.org/
To order RAND documents or to obtain additional information, contact Distribution Services: Telephone: (310) 451-7002;
Fax: (310) 451-6915; Email: order@rand.org

PREFACE

This is the final report for a Project AIR FORCE study entitled "Where Is the United States Air Force Post–Base Force, Post-BUR, and Post-QDR?" The research was sponsored by the Air Force Deputy Chief of Staff, Air and Space Operations, and was performed within the Aerospace Force Development Program of Project AIR FORCE.

The report summarizes a comparative historical review of the three major force structure reviews of the 1990s: the 1989–1990 Base Force, the 1993 Bottom-Up Review (BUR), and the 1997 Quadrennial Defense Review (QDR). It describes key assumptions, decisions, and outcomes of these reviews, focusing on elements related to strategy, forces, and resources, and summarizes key lessons learned.

The research engendered a comparative case study of the three defense reviews, where inputs (assumptions about threats, strategy, budgets) could be compared both with outputs (decisions and other outcomes of each review) and with actual implementation experience. Wherever possible, the historical inquiry was grounded in empirical data on budgets, force structure, manpower, and other issues. Despite their inherent limitations, the primary source materials for the research were official documents, including the reports, briefings, press conferences, and other outputs of the reviews themselves, as well as planning and budget documents, testimony, memoranda and other records of the reviews, and their implementation. These were supplemented by other materials as necessary. In conducting this research, the authors considered and rejected an interview-based approach to the analysis, because many defense issues (e.g., readiness) remained contentious, and the principals involved in key decisions remained active partisans in these debates. The research was conducted during fiscal year 2000, with only modest efforts made to update the material thereafter.

The report should be of interest to policymakers, planners, and others involved in the defense reviews of 2001 and to defense analysts and scholars.

Project AIR FORCE

Project AIR FORCE, a division of RAND, is the Air Force federally funded research and development center (FFRDC) for studies and analysis. It provides the Air Force with independent analyses of policy alternatives affecting the development, employment, combat readiness, and support of current and future aerospace forces. Research is performed in four programs: Aerospace Force Development; Manpower, Personnel, and Training; Resource Management; and Strategy and Doctrine.

CONTENTS

Preface iii
Figures ix
Tables xi
Summary xiii
Acknowledgments xxxi
Acronyms xxxiii

Chapter One
  INTRODUCTION 1
  Organization of This Report 2

Chapter Two
  THE BASE FORCE: FROM GLOBAL CONTAINMENT TO
     REGIONAL FORWARD PRESENCE 5
  Building the Base Force 6
    Background 6
    Strategy Under the Base Force 9
    Building the Force 15
    Resources 20
  Implementing the Base Force 23
    Strategy 23
    Force Structure and Manpower 24
    Defense Reform and Infrastructure 26
    Modernization 27
    Resources 28
    Challenges over the Horizon 34

Assessment	36
Capability to Execute the Strategy	36
Readiness	38
Modernization	38
Section Conclusions	39

Chapter Three
THE BOTTOM-UP REVIEW: REDEFINING POST–COLD WAR STRATEGY AND FORCES

THE BOTTOM-UP REVIEW: REDEFINING POST–COLD WAR STRATEGY AND FORCES	41
Building the BUR Force	42
Background	42
Strategy Under the BUR	45
Building the Force	47
Resources	56
Implementing the BUR	59
Strategy	59
Force Structure and Manpower	62
Infrastructure	65
Modernization	66
Resources	68
Assessment	75
Capability to Execute the Strategy	75
Readiness	76
Modernization	78
Section Conclusions	79

Chapter Four
THE 1997 QUADRENNIAL DEFENSE REVIEW: SEEKING TO RESTORE BALANCE

THE 1997 QUADRENNIAL DEFENSE REVIEW: SEEKING TO RESTORE BALANCE	83
Building the QDR Force	84
Background	84
Strategy Under the QDR	87
Building the Force	93
Resources	96
Implementing the QDR	100
Strategy	100
Force Structure and Manpower	102
Defense Reform and Infrastructure	104
Modernization	105
Resources	106

Assessment	112
Capability to Execute the Strategy	112
Readiness	114
Modernization	117
Section Conclusions	118
Chapter Five	
CONCLUSIONS	121
Appendix: POSTSCRIPT: FROM DEFICIT POLITICS TO THE POLITICS OF SURPLUS	127
Bibliography	135
Recommended Reading	147

FIGURES

2.1.	USAF Aircraft in Contingency Operations, 1/90–1/93	7
2.2.	The Base Force and the Spectrum of Conflict (1992 JMNA)	13
2.3.	USAF Force Structure and Manpower Reductions, FY 1990–1993	29
2.4.	"Pitchfork" Chart, Circa 1992	30
2.5.	Long-Term DoD Budget Plans, FY 1990–1991 Through FY 1994	31
2.6.	The Bow Wave in the Base Force Theater Air Program (BUR)	35
3.1.	USAF Aircraft in Contingency Operations, 1/93–10/93	43
3.2.	The BUR's View of Conflict Dynamics	51
3.3.	The BUR's Plan for Eliminating the Bow Wave in the Theater Air Program	58
3.4.	Maximum Deployment in Peace Operations, 1982–1998 (CBO)	59
3.5.	USAF Aircraft in Contingency Operations, 1993–1998	63
3.6.	USAF Force Structure and Manpower Reductions, FY 1990–1997	65
3.7.	"Pitchfork" Chart, Circa 1994	69
3.8.	Planned and Actual Procurement, FY 1994–1998 Plans	79
4.1.	The QDR's Investment Challenge	97
4.2.	Planned Procurement Under the BUR, FY 1998 Defense Program, and QDR	99

4.3.	USAF Aircraft in Contingency Operations, 5/97–8/99	101
4.4.	USAF Force Structure and Manpower Reductions, FY 1990–1999 (FY 1990 base year)	104
4.5.	DoD Plans for Budget Authority, FY 1997–2001 Budget Requests	108
4.6.	Planned and Actual BA, Air Force Procurement	111
4.7.	Planned and Actual BA for Aircraft Procurement, Air Force	112
4.8.	Planned and Actual BA for RDT&E, Air Force	113
A.1.	Defense Aggregates as a Percentage of GDP, 1940–2001	131
A.2.	Notional Budget Consequences of a "Four Percent Solution"	131

TABLES

S.1.	Proposed Force Structure Changes: Base Force, BUR, and QDR	xxviii
2.1.	Proposed Strategic Forces Package as of August 1992	17
2.2.	Proposed Base Force Conventional Force Packages as of August 1992	18
2.3.	DoD Budget Authority by Title, FY 1990 and FY 1993	22
2.4.	Planned Base Force Changes to Force Structure and Manpower, FY 1990–1997	25
2.5.	Base Force Planned vs. Actual Force Structure Changes, FY 1990–1993	26
2.6.	Bush Administration Long-Term Defense Budget Plans, FY 1991–1994 (percentage of total budget authority)	32
2.7.	Planned vs. Actual USAF Spending (BA in billions of dollars)	33
2.8.	Air Force Investment and O&S Spending, FY 1990–1993	33
3.1.	Evolution of Future Years Defense Programs in 1993 (BA in billions of dollars)	44
3.2.	Alternative Force Options Considered in the BUR	49
3.3.	Assignment of Forces in the BUR, October 1993	54
3.4.	The BUR's Long-Range Forecast for DoD	56
3.5.	DoD Budget Authority by Title	71
3.6.	Difference Between Bush and Clinton Budgets for FY 1994–1999	72
4.1.	FY 1998 Long-Range Forecast for DoD Spending	87
4.2.	Assignment of Forces in the QDR, May 1997	94

4.3.	Planned DoD Personnel End-Strength Levels, FY 1998–2003 (in thousands)	103
4.4.	FY 1999 Long-Range Forecast for DoD Spending	107
4.5.	DoD Budget Plans, FY 1998–2001 (BA in billions of dollars)	108
4.6.	DoD Procurement Plans, FY 1998–2001 (BA in billions of dollars)	109
4.7.	DoD RDT&E Plans, FY 1998–2001 (BA in billions of dollars)	110
4.8.	Air Force Spending Plans, FY 1998–2001 (BA in billions of dollars)	110
5.1.	Proposed Force Structure Changes: Base Force, BUR, and QDR	123
A.1.	Annual Deficit or Surplus, FY 1981–2000 (in billions of dollars)	128
A.2.	Caps on BA and Outlays, FY 1991–2002 (in billions of dollars)	129
A.3.	Defense and Nondefense Discretionary Outlays, FY 1991–2001 (in billions of dollars)	130
A.4.	FY 2002 President's Budget Request and "Four Percent Solution" (discretionary budget authority in billions of dollars)	132
A.5.	Comparison of Bush and Clinton FY 2002 Defense Budgets (discretionary budget authority in billions of dollars)	133

SUMMARY

The post–Cold War era—which arguably can be dated to the fall of the Berlin Wall in November 1989—has been one of immense change, and one that created equally formidable challenges for defense planners. During this period, profound transformations took place in all key elements of the policymaking environment. These included changes in the shape of the international environment, the threats to U.S. interests, and U.S. national security and military strategy. Changes also occurred in the assignment of forces, in the patterns by which forces were employed abroad, and in U.S. military force structure and personnel levels. In addition, substantial reductions were made in defense budgets. These changes—which took place at different rates and at times moved in opposing directions—placed tremendous strain both on the machinery used for deliberative planning and on the policymakers who sought to strike a balance between strategy, forces, and resources. The result was a gap that widened rather than narrowed over the decade.

This report provides contextual historical background for the defense reviews of 2001, including a Quadrennial Defense Review.[1] It focuses on how each of the three reviews conducted over the past decade addressed three key elements—strategy, forces, and resources—and describes the major assumptions, decisions, outcomes, planning, and execution associated with each.

---

[1] Section 118 of Title 10, U.S. Code, provides the statutory requirement for a Quadrennial Defense Review "during a year following a year evenly divisible by four."

THE BASE FORCE

Assumptions, Decisions, and Outcomes

The combination of favorable threat trends and adverse macroeconomic trends, including a deepening recession and the soaring budget deficit, and congressional calls for a "peace dividend" made it impossible for the new administration to protect the defense program after 1989.

In early 1989, the administration rejected the Joint Chiefs' proposal of 2 percent annual real growth and decided instead on a flat budget for one year (FY 1990) while the situation clarified, with modest real growth planned thereafter. Although it would not be until late 1990 that final budget levels were established, the Base Force and the administration's national security review in 1989 were both to be predicated on the assumption that a 25 percent reduction in force structure and a 10 percent reduction in defense resources were possible.

The revolution in the Soviet Union that had begun with General Secretary Mikhail Gorbachev's ascension led to a remarkable sequence of events, beginning with the fall of the Berlin Wall in November 1989 and culminating in the collapse of the Soviet Union in 1991—well after the public release of the Base Force.

The Base Force that was developed under Chairman of the Joint Chiefs of Staff (CJCS) Colin Powell benefited from earlier work by the Joint Staff and evolved in parallel with a larger administration review of national security and defense strategy: National Security Review 12. The aim of the Base Force was to provide a new military strategy and force structure for the post–Cold War era while setting a floor for force reductions. The floor was necessary in part to avoid creating the level of churning that might "break" the force, in part to secure the backing of the service chiefs, and later to hedge against the risks of a resurgent Soviet/Russian threat. For his part, Defense Secretary Richard Cheney's review of past defense drawdowns had animated a desire to avoid the sorts of problems wrought by the haphazard demobilizations that had followed World War II, Korea, and Vietnam. To accomplish his post–Cold War build-down in a manner that would ensure the health of the force, the secretary formed a strategic alliance with Chairman Powell that came at the price of recognizing the chairman's own constraints.

The Base Force—conceived as the minimum force necessary to defend and promote U.S. interests in the post–Cold War world—consisted of four force packages oriented toward strategic deterrence and defense (strategic forces), forward presence (Atlantic and Pacific forces), and crisis response and reinforcement (contingency forces).

The size of the force was to be determined primarily by regional needs and not on the basis of its capability to fight multiple major theater wars (MTWs); while the main conventional threat for the Base Force was the potential for major regional conflicts involving large-scale, mechanized cross-border aggression, the multiple-MTW construct that was to dominate defense planning for the remainder of the decade was an afterthought. The two-conflict case was simply one of a number of illustrative planning scenarios that were developed after the threat-based planning environment collapsed to test the capabilities of the force. In fact, the two-simultaneous-conflict scenario was a case that Chairman Powell testified would put the Base Force "at the breaking point." Although flexible general-purpose forces were needed to address the entire "spectrum of threat," from humanitarian assistance and noncombatant evacuation operations to major regional conflicts, there is little evidence that substantial involvement in peacekeeping and other peace operations was anticipated during the development of the Base Force.

In 1990, as the Base Force was being finalized, pressures for defense spending reductions were given additional impetus by the federal deficit, which had ballooned in the final years of the 1980s, and by the possibility that crippling spending cuts would automatically be triggered under the Gramm-Rudman-Hollings antideficit laws. In the June 1990 budget summit, Secretary Cheney used the Base Force to illustrate the feasibility of a 25 percent smaller force that could provide approximately 10 percent in defense cuts. It was not until the October 1990 budget summit, however, that deficit and discretionary spending caps were finalized, resulting in deeper-than-expected cuts to defense budgets. The Base Force was presented with the next President's Budget submission in early 1991.

Throughout the Base Force deliberations, the Air Force's principal aim was to preserve its modernization and acquisition programs while ensuring that the pace of reduction did not harm people or the future quality of the force. Accordingly, early in the process of defining the force, Air Force leaders accepted both the Base Force concept and the implication that USAF force structure would be further reduced, thus showing a willingness to trade force structure to maintain modernization. The number of tactical fighter wing equivalents would be reduced by about one-third, with most of the reductions coming from the active force. Strategic long-range bombers were also to be reduced, while the conventional capabilities of the bomber force were to be improved.

Planning and Execution

Although the Base Force set force structure targets for 1995 and 1997, the outcome of the 1992 presidential elections meant that the Base Force was actually implemented over only two years, FY 1992–1993.

The Base Force's effort to adapt conventional forces to the post–Cold War world resulted both in force reductions and in modest changes to the allocation of resources among the services. These changes suggested a declining emphasis on land forces (the Army share of DoD budget authority fell from 26.8 percent in 1990 to 24.3 percent in 1993) and an increasing emphasis on aerospace power (with the Air Force share rising from 31.7 percent in 1990 to 32.9 percent in 1993).

While few problems were encountered in realizing the planned 25 percent force structure and 20 percent active manpower reductions, policymakers had difficulty realizing the 20 percent reductions to reserve-component manpower, particularly to Army and Marine Corps reserves. This resulted in higher-than-anticipated reserve-component manpower levels and lower-than-expected savings. In the end, the rate at which civilian manpower fell over 1990–1993 was greater than that for active- or reserve-component personnel and greater than had initially been planned.

Notwithstanding Air Force leaders' hopes to trade force structure for modernization and acquisition, greater-than-expected budget cuts led to the reduction or termination of a number of high-priority Air

Force modernization programs during the course of—or as a result of—the Base Force, including the B-2 (from 132 to 75 aircraft, and subsequently to 20 aircraft), the F-22 (from some 750 to 648 aircraft), and the C-17 (from 210 to 120 aircraft). Nevertheless, efforts to improve the conventional capabilities of long-range bombers and to expand capabilities for precision-guided munitions were begun as a result of the 1992 Bomber Roadmap and other initiatives.

To provide additional savings, the administration also pursued defense reform and infrastructure reductions through the Defense Management Review (DMR) and the Base Realignment and Closure (BRAC) Commission. The 1989 BRAC round identified 40 bases for closure, and the 1991 round envisioned closing another 50, for recurring annual savings of perhaps $2.5 billion to $3.0 billion a year. Nevertheless, by late 1992–1993, concerns had arisen that not all of the anticipated $70 billion in savings from the DMR and BRAC rounds would be realized—a problem that the early Clinton administration faced. In fact, the percentage of USAF total obligational authority devoted to infrastructure increased from 42 to 44 percent in 1990–1993.

In addition to the difficulties encountered in reducing reserve-component military personnel and realizing savings from infrastructure reductions and defense reform, defense budgets continued to decline; each of the Bush administration's budget requests from FY 1990 to FY 1994 envisioned lower spending levels.[2] In the longer term, the problems were even more challenging. By December 1991, for example, the Congressional Budget Office was projecting that the Base Force could not be maintained and modernized in the 1993–1997 period if Congress and the administration complied with the limits of the 1990 Budget Enforcement Act and that beyond 1997, shortfalls of $20 billion to $65 billion could be expected as policymakers sought to carry out necessary modernization.

Testimony suggests that Base Force policymakers expected that they would need to face these problems after 1992: Chairman Powell suggested that by 1995 a new Base Force, engendering additional cuts to force structure, might be necessary, and Secretary Cheney suggested

---

[2]The Bush administration never formally submitted its FY 1994 budget request.

that increases in real defense spending might be needed in the out years to cover modernization needs, including an anticipated procurement "bow wave."

THE BOTTOM-UP REVIEW

Assumptions, Decisions, and Outcomes

The 1993 *Report on the Bottom-Up Review* (BUR) was the second major force structure review of the decade that aimed to define a defense strategy, forces, and resources appropriate to the post–Cold War era.

Under the BUR, force structure and manpower reductions would accelerate and would surpass those planned in the "Cold War-minus"-sized Base Force, leading to a total reduction in forces of roughly one-third—a level well beyond the Base Force's planned 25 percent reduction, most of which had already been achieved by the end of FY 1993. Budgets would also fall beyond planned Base Force levels as a result of the BUR. Indeed, it appears that budget top lines were established before either force structure or strategy had been decided.

The aim of the BUR was to provide "a comprehensive review of the nation's defense strategy, force structure, modernization, infrastructure, and foundations." While embracing the Base Force's regionally focused strategy and emphasis on strategic deterrence, forward presence, and crisis response, the BUR redefined the meaning of engagement, giving increased rhetorical and policy importance to U.S. participation in multilateral peace and humanitarian operations and setting the stage for an increased operational tempo and rate of deployment, even as force and budgetary reductions continued.

During the 1992 presidential campaign, candidate Bill Clinton had argued that changes in the threat environment, taken together with the nation's poor economic circumstances, made possible a cut of approximately $60 billion in defense spending. By the time the FY 1994 budget was submitted in February 1993, the administration was planning force structure reductions to meet savings goals of $76 billion over FY 1994–1997 and $112 billion over FY 1994–1998; the $104 billion in cuts envisaged in the October 1993 BUR were only slightly smaller than those documented six months earlier in the President's

Budget. Put another way, the cuts to the defense top line planned in the FY 1994 budget were, within a few billion dollars in any given year, identical to those in the FY 1995 budget request that implemented the BUR. As a result, the strategy, force structure, modernization, and other initiatives described in the BUR were to be driven as much by the availability of resources as by the threats and opportunities in the emerging international environment and the administration's own normative foreign policy aims.

In carrying out the budget cuts, Clinton administration policymakers hoped to reduce defense spending without raising questions about their commitment to the nation's defense. The result was more modest cuts in force structure in the BUR than had been advocated by then–House Armed Services Committee Chairman Les Aspin in 1992, but deeper cuts in defense resources than had been advocated during the campaign. Another result was that a strategy was ultimately overlaid on a force structure that was justified in warfighting terms but would soon became preoccupied instead with operations in support of the administration's still-crystallizing strategy of "engagement and enlargement."

As described in the BUR, four "strategies" were considered, each of which identified a plausible mix of operations that future U.S. forces might need to conduct. After choosing—and then rejecting—the second strategy and force structure ("win-hold-win"), the BUR chose the force associated with a newly developed third strategy: the capability to win two nearly simultaneous major regional conflicts (MRCs). This force differed only slightly from the win-hold-win force but provided some additional capabilities for carrier-based naval presence operations, enhanced-readiness Army reserve-component forces, and a number of additional force enhancements (e.g., precision attack, strategic mobility) that aimed to improve the force's ability to underwrite two conflicts with a smaller force structure than the Base Force. This force was also smaller than that associated with a fourth strategy, which provided a capability for winning two nearly simultaneous MRCs *plus* conducting smaller operations. As was the case in the Base Force, the BUR force structure in part reflected Chairman Powell's negotiations with the service chiefs over force levels. Although the CJCS went to great pains to emphasize that the BUR force was designed for warfighting, the BUR also anticipated a high level of commitment to peace, humanitarian, and other operations. Accordingly, it laid down an elaborate logic to ensure the force's ability to disengage from peacetime operations and established several management oversight groups to monitor readiness and other risks that might result if this ambitious strategy were to be executed with smaller forces.

In the face of the additional anticipated budget cuts, the BUR undertook only selective modernization and generally sought to address key problems such as the "bow wave" in the theater air program. The BUR also supported several so-called new initiatives that were directed toward improving U.S. capabilities in areas other than traditional warfighting. As described above, the BUR reported that it could support the strategy and force structure while realizing $104 billion in savings from the Bush baseline in nominal dollars; Office of the Secretary of Defense policymakers, however, are reported privately to have expected only some $17 billion in savings.

As it had with the Base Force, the Air Force embraced the new strategy and its emphasis on long-range aerospace power, including long-range conventional bombers, strategic mobility, enhanced surveillance and targeting, and precision-guided attack—even as Air Force leaders expressed disappointment that the BUR would fail to affect roles and missions and would instead result in a force they described as "Cold War-minus-minus." And as with the Base Force, the Air Force sought to trade force structure and end strength for continued modernization. Although Air Force commitments to contingency operations had already increased by the time of the BUR, the Air Force does not appear to have pressed the case that peacetime presence and contingency operations should also be considered in sizing the USAF—an argument that the Navy had profitably used to justify a 12-carrier force.

Planning and Execution

The BUR was implemented over four years via the FY 1995–1998 budget submissions.

Although the nominal strategy and force structure chosen in the BUR was predicated largely on the ability to fight and win two nearly simultaneous wars, it also implied higher levels of involvement in peace, humanitarian, and other smaller-scale operations than had the Base Force. The new strategy's heavy emphasis on peace operations resulted in commitments throughout the 1993–1998 period that from a historical perspective were frequent, large, and of long duration. The evidence suggests that policymakers—including those in the Air Force—underestimated these demands, which, by some accounts, eventually amounted to the equivalent of one MTW's worth of forces. The result was growing congressional and other concern about the potential impact on warfighting of smaller-scale contingencies (SSCs).

The new strategy placed unprecedented demands on the Air Force in servicing peacetime contingency operations over this period while remaining ready for warfighting. While the number of Air Force aircraft deployed to contingency operations in early 1990 had been nominal, it increased dramatically with the Iraqi invasion of Kuwait and the Gulf War in 1990–1991 and then remained at a substantially higher level than before the war as a result of the need to sustain operations in northern and southern Iraq. Modest increases in the number of aircraft in contingency operations were seen thereafter as additional commitments accumulated, particularly in the Balkans. At any given time, more than 200 USAF aircraft would typically be deployed throughout the 1993–1998 period, although occasional peaks of 250 to 350 aircraft were also seen. With the force structure decisions taken in the BUR, however, the die had already been cast, resulting in a smaller force underwriting a more ambitious strategy and a fourfold increase in operational tempo over that prior to the fall of the Berlin Wall.

Although force structure goals were achieved relatively quickly, infrastructure reductions lagged; for example, the percentage of USAF total obligational authority devoted to infrastructure fell from 44 percent in 1993 to 42 percent in 1998. And although the incremental costs of peace operations were only about $14.1 billion over FY 1994–1998, the actual costs of the defense program turned out to be much higher than anticipated in the FY 1994 budget, the BUR, or the FY 1995 budget that implemented the BUR.[3] As a result of the imbal-

---

[3]Office of the Secretary of Defense Comptroller data were provided by the Joint Chiefs of Staff and include contingency operations in Southwest Asia, Bosnia, Haiti, Cuba, Rwanda, Somalia, and Kosovo. The incremental costs over the FY 1994–1999 period were $20.1 billion.

ance between resources on the one hand and strategy and forces on the other, only some $15 billion of the $104 billion in anticipated savings reported by the BUR were realized—a level not significantly different from the $17 billion in savings that was privately said to be expected. Part of the difference was made up through emergency supplemental appropriations and by gradually increasing subsequent budgets to try to close the gap.

The resulting gaps had two principal results. First, despite the high priority and high levels of spending on operations and support (O&S) accounts, readiness problems emerged, many of them resource-related, while the risks associated with executing the national military strategy grew. Second, over the 1995–1997 period, spending on modernization fell well below the levels planned in the FY 1994 (transition) and 1995 (BUR) budgets; instead, funds routinely "migrated" from investment accounts to O&S accounts, resulting in program stretch-outs and delays to planned modernization efforts.

In retrospect, then, it appears that the force chosen by the BUR was less suitable to the high levels of peacetime engagement in contingency operations that were actually observed in subsequent years than the force deemed capable of winning two nearly simultaneous MRCs *plus* conducting smaller operations. Furthermore, the failure to achieve most of the anticipated savings reported by the BUR suggests that the BUR force in fact required a Base Force–sized budget. In the end, the mismatch between a more ambitious strategy of engagement and the forces and resources that were declining at different rates made it impossible for the services to support the dual priorities of readiness and modernization during the years in which the BUR was implemented.

THE QUADRENNIAL DEFENSE REVIEW

Assumptions, Decisions, and Outcomes

The 1997 *Report of the Quadrennial Defense Review* (QDR) considered the potential threats, strategy, force structure, readiness posture, military modernization programs, defense infrastructure, and other elements of the defense program needed for the 1997–2015 time frame and beyond.

The QDR was intended to provide a blueprint for a strategy-based, balanced, and affordable defense program. Lingering concerns about the deficit and the austere budgetary environment that resulted, however, placed continued constraints on defense resources, leading to the assumption of flat, $250 billion-a-year defense budgets. Equally important, the QDR aimed—within a flat budget and with only modest adjustments to force structure—to rebalance the defense program and budget to address some of the key problems that had developed during the BUR years, including the adverse effect of SSCs and the "migration" of funds from modernization (particularly procurement) accounts to operations accounts. The combination of budgetary constraints, Defense Secretary William Cohen's outsider status as the newest member of (and sole Republican in) the Clinton cabinet, and the dominant influence of the services in the review appear to have made it a foregone conclusion that the QDR would fail to challenge the status quo and would fall short of achieving the balance that was sought.

The QDR generally accepted the normative and other underpinnings of the BUR's strategy, reaffirmed the BUR's emphasis on two nearly simultaneous MTWs as the principal basis for force sizing, and posited that the United States might have to fight one or two MTWs over the 1997–2015 period. It also anticipated continued involvement over the same period in the kinds of SSCs that had been described in the BUR, including peace and humanitarian operations.

The QDR did make several important adjustments to the BUR strategy, however, two of which had substantive importance. First, it placed increased emphasis on the halt phase in MTWs. Second, it gave increased rhetorical recognition to the demands of SSCs and recognized the potential need to respond to multiple concurrent SSCs. Yet while it aimed to provide "strategic agility"—i.e., the capability to transition from global peacetime engagement to warfighting—the QDR did not advocate significant adjustments in force structure or resourcing to accommodate these demands. Finally, the QDR articulated a somewhat more cautious and nuanced employment doctrine than had the BUR, distinguishing among situations involving vital, important but not vital, and humanitarian interests and identifying the sorts of responses appropriate to each.

The QDR rejected two straw men—a U.S. strategy of isolationism and one in which the United States would serve as "world policeman"—in favor of a strategy of engagement and a path that balanced current demands against an uncertain future. The result of the assessment was promoted as a more balanced strategy—dubbed "shape, respond, and prepare now"—that embraced both active engagement and crisis response options while also advocating increased resources for force modernization.

The QDR rejected a 10 percent cut in force structure because it would result in unacceptable risk, presumably both to warfighting capability and to the force's ability to engage in SSC operations. Accordingly, changes to force structure involved only modest reductions as well as some restructuring. Among the most important of these changes was the decision to move one Air Force tactical fighter wing from the active to the reserve component, leaving slightly more than 20 tactical fighter wings in the force structure.

With force structure cuts essentially off the table, savings were to be achieved through manpower cuts. Secretary Cohen instructed the services to cut the equivalent of 150,000 active military personnel to provide $4 billion to $6 billion in recurring savings by FY 2003; the QDR reported the decision to further reduce active forces by 60,000, reserve forces by 55,000, and civilians by 80,000 personnel.

Finally, in a bow to the procurement spending goal that CJCS John Shalikashvili established in 1995, the QDR made a long-term commitment to achieve $60 billion a year in procurement spending by 2001—a nominal level of procurement spending that was in fact less than what the chairman had originally specified.[4] Nevertheless, the QDR's modernization effort reflected the same response to the tight budgetary environment as the BUR—namely, to fund only "selective" modernization. In other words, to make the program affordable, the QDR made additional cuts to a number of acquisition

---

[4]In 1995, Chairman Shalikashvili had called for the annual procurement budget to reach $60 billion by FY 1998—the minimum needed level of procurement as determined by the Defense Program Projection, and originally conceived in terms of constant FY 1993 dollars. By the time of the QDR in 1997, the $60 billion had become a nominal dollar target that, accordingly, had lost some of its earlier purchasing power—all the more so as a result of its deferral from FY 1998 to FY 2001.

programs and advocated additional savings through further infrastructure reduction and defense management reform.

Nevertheless, in many respects the QDR presented the Air Force with important opportunities to promote the concepts and core competencies developed in its most recent strategic planning exercise. For example, the review emphasized rapid response and an early, decisive halt of cross-border aggression, which played to Air Force strengths in long-range precision strike and mobility. Similarly, the QDR's focus on reducing the stresses created by SSCs accented a number of post-BUR USAF innovations, such as the air expeditionary force concept. Finally, the QDR strategy's rhetorical emphasis on preparing for an uncertain future played to a long-standing Air Force priority: investment in advanced technologies. However, Air Force leaders' hopes to use the QDR to challenge the status quo and to transform U.S. forces were reportedly dashed by CJCS Shalikashvili's message that there would be no "Billy Mitchells" in the QDR.

Planning and Execution

The QDR, together with its claim to have successfully balanced the defense program, was met with some skepticism both by Congress and by many other observers. Nevertheless, rather than adjusting discretionary caps, the administration and Congress continued their previous pattern of reducing the resource gap through emergency supplementals and year-to-year increases.

In some respects, the new strategy elements of shaping and responding differed little from the BUR's strategy of engagement and enlargement: Both relied heavily on forward presence and crisis response capabilities, and both were concerned with ensuring stability in the near term in regions of vital interest. And although the QDR had anticipated continued participation in SSCs, actual U.S. participation in peace and other contingency operations turned out to be somewhat higher than anticipated. In February 1998, for example, CJCS Henry Shelton reported that 1997 had seen 20 major operations and many smaller ones, with an average of 43,000 service members per month participating in contingency operations.

Given the modest changes to force structure recommended by the QDR, it should come as little surprise that with only a few exceptions, force structure changes for major force elements were already in place in the FY 2001 President's Budget and defense program.

While the other services were expected to hit the QDR manpower targets by 2003, the manpower reductions programmed for the Air Force in the FY 1999 budget suggested that the Air Force would not achieve its targets by this date. The Air Force aimed to achieve its manpower reductions principally through aggressive competitive outsourcing of certain functions, the restructuring of combat forces, and the streamlining of headquarters. These plans encountered difficulties in their execution, however, resulting in a smaller personnel reduction than had been identified in the QDR and in a failure to realize all of the anticipated savings.

Evidence of readiness problems continued to accumulate in the wake of the QDR to the point at which, in the fall of 1998, the service chiefs reported that readiness problems were both more prevalent and more serious than had earlier been reported. The risks associated with executing the two-conflict strategy also increased over this period, with the risk associated with the second conflict now reported to be high, seemingly as a result of lower readiness levels for forces earmarked for the second MTW and shortfalls in strategic mobility. The result of these developments, which played out in late 1998 and early 1999, was a FY 2000 budget request that entailed the first real increase in defense resources in more than a decade: Approximately $112 billion in additional resources, primarily to address readiness problems, was committed in the FY 2000 Future Years Defense Program (FYDP), nearly restoring the funding that had been taken out in 1993–1994.

Although the QDR sought to meet its $60 billion modernization goal by reducing excess facilities, closing additional bases, and realigning and streamlining infrastructure, additional rounds of BRAC were not authorized by Congress, and savings from defense reform efforts, while not insubstantial, were disappointing. The QDR's modest modernization goals appear to have been met, although the $60 billion target falls well short of the estimated $80 billion to $90 billion or more that is believed to be needed for recapitalization; funding for transformation of the force was even less generous and fell well below the $5 billion to $10 billion recommended by the National Defense Panel, which critiqued the QDR.

CONCLUSIONS

This history of the three major defense strategy reviews of the past decade aims to highlight the inputs (assumptions, threats, and domestic environments), outputs (decisions and other outcomes), and implementation experience of each review. After identifying some common features of the reviews, we offer some lessons regarding strategy, forces, and budgets and then close with some thoughts on how defense planning might be improved.

Stepping back from their details, the reviews appear to have shared at least three main features, each of which could benefit from additional scrutiny:

- First, each assumed that the most important (and taxing) mission for conventional forces was halting and reversing cross-border aggression by massed mechanized forces. Post–Cold War and post–Gulf War reductions in mechanized forces abroad and the war in Kosovo suggest, however, that fighting dispersed forces in uneven terrain may be even more problematic cases in the future.

- Second, each review in its own way treated presence and smaller-scale peace and other contingency operations as "lesser-included cases" that could successfully be managed by a force structure designed primarily for warfighting—and assumed that these contingency operations would impose minimal costs and risks for warfighting. SSCs have not been lesser-included cases, however, and have instead represented competing claimants for increasingly scarce defense resources.

- Third, each review suffered from the absence of a bipartisan consensus on a post–Cold War foreign and defense policy, and this made the gaps that emerged between strategy, forces, and budgets particularly salient while arguably impeding their successful resolution. The new administration should consider how best to establish a shared vision of the nation's defense priorities, a better partnership with Congress, and a process for fuller consideration of defense funding needs.

We now turn to some lessons for strategy, forces, and budgets.

Regarding *strategy*, the historical record suggests that it is critically important to understand that changes in strategy—a regular feature of presidential transitions and defense reviews—can have a range of important ramifications. The change in normative aims and conception of engagement pursued by the Clinton administration and documented in the BUR, for example, underscored the importance of ethnic conflict and civil strife, promoted peace operations as a more important tool of U.S. policy, and had strong implications for the resulting pattern of U.S. force employment. Having failed to fit force structure and budgets to strategy, the resulting impacts could and should have been better anticipated and resources realigned to mitigate or eliminate them.

Another critical result has to do with *force structure*. Table S.1 shows that while there have been substantial reductions in force structure

Table S.1

Proposed Force Structure Changes: Base Force, BUR, and QDR

Service[a]	FY 1990	1997 Base Force	1999 BUR Force	2003 QDR Force	FY 2001
Air Force					
TFWs (AC/RC)	24/12	15.3/11.3	13/7	12+/8	12+/7+
Bombers (active)	228	181	184	187	181
Land-based ICBMs	1000	550	550	550	550
Navy					
Aircraft carriers	15/1	12/1	11/1	11/1	12/0
Battle force ships	546	448	346	306	317
Marine Corps					
Divisions (AC/RC)	3/1	3/1	3/1	3/1	3/1
Army					
Divisions (AC/RC)	18/10	12/8[b]	10/5+	10/8	10/8
End strength					
Active duty	2070	1626	1418	1360	1382
Reserve	1128	920	893	835	864

[a]TFWs = tactical fighter wings; AC = active component; RC = reserve component; ICBM = intercontinental ballistic missile.
[b]RC includes two cadre divisions.

and manpower, only a modest amount of force reshaping has actually taken place. Efforts to meaningfully modernize and transform the force have been hampered by a high discount rate that has elevated current-day threats, force structure, and readiness concerns while effectively discounting longer-term needs. The irony, of course, is that readiness has also suffered.

With respect to *budgets*, there seems to have been a chronic reluctance to acknowledge what reasonable-risk versions of a strategy and force structure might really cost. While gaps between strategy, force structure, and resources are not unprecedented,[5] the tacit agreement of the executive and legislative branches to avoid debates over issues of strategy and policy may actually have impeded full disclosure and consideration of the problems that plagued the defense program for much of the decade. Instead, the reliance on modest year-to-year revisions that did not upset discretionary spending limits, coupled with the recurring exploitation of the loophole provided by emergency supplementals to mitigate particularly acute shortfalls, meant that the debates would occur only at the margin. Failure to tackle these issues head on may have retarded the recognition and remediation of the growing gaps between strategy, forces, and resources.

Turning to the Air Force, the USAF arguably did quite well in recognizing what it could contribute in the post–Cold War environment and was trained and equipped to support the operations it was called on to undertake. Nevertheless, it did less well recognizing emerging opportunities and challenges, recognizing events whose outcomes it did not control, and positioning itself best in those circumstances. It also failed to successfully make the case for a more probing DoD examination of how future roles and missions—and budgets—might be adjusted to better meet the needs of the emerging environment.

---

[5]During the Cold War period, for example, airlift capacity remained well short of the 66 million ton-miles per day (MTM/D) that was the stated requirement for responding to a Soviet–Warsaw Pact attack across the inter-German border and a Soviet invasion of Iran. Current military airlift surge capacity is judged to be nearly 20 percent short of the requirement established by the 1995 Mobility Requirements Study Bottom-Up Review Update and roughly 23 to 31 percent short of the requirement established in the more recent MRS05.

Shifting to the present, the new administration's defense review will wrestle with the same questions its predecessors faced: What are to be the nation's aims in the world? What are the main threats and opportunities it faces? What strategy and force structure will best meet the needs of the nation? What resources are needed to ensure low to moderate execution risk in that strategy, and capable and ready forces, both now and over the next 20 to 30 years?

In answering these questions, the Department of Defense—and the Air Force—would profit from an assumption-based planning approach in which signposts are established that can be used to gauge whether the key assumptions on which planning is predicated are still justified. These include assumptions about future threats, the likely frequency and mix of future missions, the adequacy of forces to undertake these missions, and what resources are needed.

Such a planning approach is desirable by virtue of another great lesson of the past decade: that failure to recognize and respond promptly and effectively to emerging gaps and shortfalls can lead to the greatest and most protracted imbalances among strategy, forces, and resources.

ACKNOWLEDGMENTS

The authors would like to thank Colonel Bruce Wong (SAF/XO/QRY) and Major Dennis Armstrong for their assistance as project monitors and Brigadier General David Deptula and Brigadier General Ronald Bath for their comments on our initial findings from this research. In addition, we wish to thank RAND colleagues David Chu and Frank Lacroix for their thoughtful reviews as well as Natalie Crawford, Tim Bonds, Carl Dahlman, Jim Dewar, Myron Hura, Gustav Lindstrom, Judy Mele, David Ochmanek, David Owen, Jed Peters, Bob Roll, Mike Scheiern, James Schneider, Tim Smith, and Donald Stevens of RAND for their assistance in this project. We are also grateful to Andrea Fellows for her editing of the manuscript. Finally, we wish to thank Steven Daggett of the Congressional Research Service for providing data comparing the BUR's planned savings with actual savings over its implementation.

ACRONYMS

AC	Active component
AEF	Aerospace expeditionary force
AFRES	Air Force Reserve
AGS	Air Ground Surveillance
ALCM	Air-launched cruise missile
ANG	Air National Guard
APS	Afloat prepositioning ship
ARG	Amphibious Ready Group
ARNG	Army National Guard
AWACS	Airborne Warning and Control System
BA	Budget authority
BAI	Backup aircraft inventory
BRAC	Base realignment and closure
BUR	Bottom-Up Review
C^3	Command, control, and communications
C^4ISR	Command, control, communications, computers, intelligence, surveillance, and reconnaissance
CBO	Congressional Budget Office
CJCS	Chairman of the Joint Chiefs of Staff
CONUS	Continental United States
CVBG	Carrier Battle Group
DMR	Defense Management Review
DMRD	Defense Management Review Decision
DPG	Defense Planning Guidance
FEWS	Follow-on Early Warning System
FYDP	Future Years Defense Program
GPALS	Global Protection Against Limited Strikes

HA/DR	Humanitarian assistance/disaster relief
IADS	Integrated air defense system
ICBM	Intercontinental ballistic missile
IOC	Initial operational capability
IPS	Illustrative planning scenario
ISR	Intelligence, surveillance, and reconnaissance
J-5	JCS Strategic Plans and Policy Directorate
J-8	JCS Force Structure, Resources, and Assessment Directorate
JASSM	Joint Air-to-Surface Standoff Missile
JAST	Joint Advanced Strike Technology
JCS	Joint Chiefs of Staff
JDAM	Joint Direct Attack Munition
JMNA	Joint Military Net Assessment
JPATS	Joint Primary Aircraft Trainer System
JSOW	Joint Standoff Weapon
JSTARS	Joint Surveillance Target Acquisition Radar System
LD/HD	Low density/high demand
LMSR	Large, medium-speed roll-on/roll-off
LRC	Lesser regional conflict
MEB	Marine Expeditionary Brigade
MEF	Marine Expeditionary Force
MIDEASTFOR	Middle East Force
MOOTW	Military operations other than war
MPS	Maritime prepositioning squadron
MRC	Major regional conflict
MRF	Multirole Fighter
MRS	Mobility Requirements Study
MRS BURU	MRS Bottom-Up Review Update
MTM/D	Million ton-miles per day
MTW	Major theater war
NATO	North Atlantic Treaty Organization
NDP	National Defense Panel
NEO	Noncombatant evacuation operation
NMS	National Military Strategy
NSC	National Security Council
NSR	National Security Review
O&M	Operations and maintenance
O&S	Operations and support

OCOTF	Overseas Contingency Operations Transfer Fund
OMB	Office of Management and Budget
OSD	Office of the Secretary of Defense
PAA	Primary Aircraft Authorization
PACOM	Pacific Command
PPBS	Planning, Programming, and Budgeting System
QDR	Quadrennial Defense Review
RC	Reserve component
RDT&E	Research, Development, Test, and Evaluation
RMA	Revolution in Military Affairs
SAR	Selected Acquisition Report
SEAD	Suppression of Enemy Air Defenses
SECDEF	Defense Secretary
SSBN	Strategic ballistic missile submarine
SSC	Smaller-scale contingency
START	Strategic Arms Reduction Talks
TAI	Total aircraft inventory
TFW	Tactical fighter wing
TFWE	Tactical fighter wing equivalent
THAAD	Theater high-altitude area defense
TOA	Total obligational authority
UNSC	United Nations Security Council
USMC	U.S. Marine Corps
WCMD	Wind-Corrected Munition Dispenser
WMD	Weapons of mass destruction

Chapter One
INTRODUCTION

This report describes the challenges policymakers and defense planners faced in the first decade after the Cold War. It does so through the lens of an assessment of the three major force structure reviews that took place in the 1990s: the 1989–1990 Base Force, the 1993 Bottom-Up Review (BUR), and the 1997 Quadrennial Defense Review (QDR).

The post–Cold War era—which arguably can be dated to the fall of the Berlin Wall in November 1989—has been one of immense change, and one that created equally formidable challenges for defense planners. During this period, profound transformations took place in all key elements of the policymaking environment, including the shape of the international environment, the threats to U.S. interests, and U.S. national security and military strategy. Changes also occurred in the assignment of forces, in the patterns by which forces were employed abroad, and in U.S. military force structure and personnel levels. In addition, substantial reductions were made in defense budgets. These concurrent changes—which occurred at different rates and at times moved in opposing directions—placed tremendous strain both on the machinery used for deliberative planning and on the policymakers who sought to strike a balance between strategy, forces, and resources.

This report is a work of both history and policy analysis that aims at providing contextual background for the QDR in 2001. It focuses on three key elements—strategy, forces, and resources—and compares the key assumptions, decisions, outcomes, planning, and execution of the three reviews.

The assessment was guided by the following questions:

- What was the state of the world at the time of each force structure review?
- What was the U.S. military (and especially Air Force) posture going into each review?
- What were the major assumptions, conclusions, and outcomes of each review?
- What was the subsequent postreview experience?
- What lessons can be drawn from the past for the next QDR?

ORGANIZATION OF THIS REPORT

This report is organized by review, with one chapter devoted to each review. Each chapter first describes the assumptions, decisions, and outcomes of the review in question. Assumptions include beliefs about the threats and opportunities in the environment, the shape of future presence operations, and the probable missions assigned to U.S. military forces. Decisions and outcomes include the assignment of forces to underwrite the strategy, force structure goals, and priorities for the allocation of resources. Each chapter then describes the planning and execution of the decisions and outcomes of the reviews. Accordingly, Chapter Two focuses on the 1989–1990 Base Force; Chapter Three assesses the 1993 BUR; and Chapter Four discusses the 1997 QDR.

In Chapter Five, we summarize what we believe are some of the key lessons drawn from our historical analysis of the last decade's defense reviews, identify some common features of these reviews, and offer some general conclusions about how defense planning can avoid some of the problems that were encountered over the past decade.

The major force structure reviews discussed in this report took place independently of, but were influenced by, a number of other policy-relevant developments over the decade. For example, strategic nuclear and mobility forces, which were not considered in detail in the reviews, were addressed in separate, more detailed studies. Additionally, the base realignment and closure (BRAC) process for reducing infrastructure and other efforts to reform defense management were pursued to free up resources. Finally, several commissions addressed issues such as roles and missions and future strategy and force needs. The present report concentrates on the major force structure reviews and discusses these other influences only when additional context is necessary.

Chapter Two

THE BASE FORCE: FROM GLOBAL CONTAINMENT TO REGIONAL FORWARD PRESENCE

The changes to strategy and force structure that were developed under the Base Force were designed to meet the defense needs of the post–Cold War era by replacing Cold War strategy, which had focused on deterrence of Soviet aggression and had relied on forward defense, with a new strategy focused on regional threats and forward presence. The Base Force called for substantial changes in U.S. military forces, including a 25 percent reduction in force structure, an approximately 10 percent reduction in budget authority, and more than a 20 percent reduction in manpower relative to FY 1990. This chapter provides a synopsis of the key components of the Base Force in terms of strategy, force structure, and resources, and it then assesses the planning and execution of some of the key elements of the Base Force through FY 1993, the final year in which the Base Force was actually implemented.[1]

---

[1]Descriptions of the Base Force at various levels of detail can be found in Colin L. Powell, "Building the Base Force: National Security for the 1990s and Beyond," annotated briefing, September 1990; congressional testimony by Chairman Powell and Secretary Cheney in 1991 and 1992; Joint Chiefs of Staff, *Joint Military Net Assessment*, Washington, D.C., 1991 and 1992; and Joint Chiefs of Staff, *1992 National Military Strategy*, Washington, D.C., January 1992. Detailed histories of the Base Force can be found in Don M. Snider, *Strategy, Forces and Budgets: Dominant Influences in Executive Decision Making, Post–Cold War, 1989–91*, Carlisle Barracks, PA: Strategic Studies Institute, Professional Readings in Military Strategy No. 8, February 1993; Lorna S. Jaffe, *The Development of the Base Force, 1989–1992*, Washington, D.C.: Joint History Office, Office of the Chairman of the Joint Chiefs of Staff, July 1993; and Leslie Lewis, C. Robert Roll, and John D. Mayer, *Assessing the Structure and Mix of Future Active and Reserve Forces: Assessment of Policies and Practices for Implementing the Total Force Policy*, Santa Monica: RAND, MR-133-OSD, 1992. An assessment of the Base Force can be found in U.S. General Accounting Office, *Force Structure: Issues*

BUILDING THE BASE FORCE

Background

The World Situation. The period in which the Base Force emerged—from the Joint Staff's initial work in the summer of 1989 to the approval of the Base Force in the late fall of 1990 and its presentation in the spring of 1991—was clearly a tumultuous one.

The changes in the Soviet Union that were initiated with General Secretary Mikhail Gorbachev's ascension led to a dizzying sequence of events that began with the fall of the Berlin Wall in November 1989 and the reunification of Germany in October 1990. Immediately following the presentation of the Base Force in the early spring of 1991, the Warsaw Pact military structure dissolved in April 1991, the START treaty was signed in July 1991, and the Soviet Union dissolved in December 1991.

Even as traditional threats were evaporating, however, new ones were emerging outside the European theater. During the Base Force period, U.S. military forces were called on to intervene in Panama, respond to Iraq's invasion of Kuwait, and manage the consequences of refugees fleeing the instability in Haiti. In the same time frame, U.S. forces also participated in noncombatant evacuation operations in Liberia and Somalia, large-scale evacuation operations following the eruption of Mount Pinatubo in the Philippines, and large-scale humanitarian relief operations in northern Iraq and Somalia.

U.S. military operations in the early 1990s required the deployment of an increasing number of USAF aircraft. Figure 2.1 shows the number of Air Force aircraft involved in contingency operations between January 1990 and January 1993, when the Bush administration left office.[2] Prior to the war (from January 1990 until the Iraqi invasion of Kuwait in August 1990), the typical pattern of deployment to contingency operations was nominal, usually involving fewer than

---

Involving the Base Force, Washington, D.C., GAO/NSIAD-93-65, January 1993. The issue of the Base Force's treatment of Total Force policy can be found in Bernard Rostker et al., *Assessing the Structure and Mix of Future Active and Reserve Forces: Final Report to the Secretary of Defense*, Santa Monica: RAND, MR-140-1-OSD, 1992.

[2]Data are from DFI International, Washington, D.C.

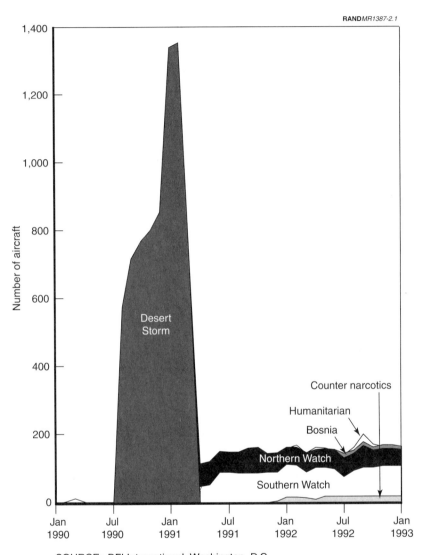

SOURCE: DFI International, Washington, D.C.

Figure 2.1—USAF Aircraft in Contingency Operations, 1/90–1/93

ten aircraft.[3] With the Iraqi invasion and the Gulf War, however, substantial numbers of USAF aircraft were deployed (and employed). Although these numbers declined after the Gulf War, they remained substantially higher than during the prewar period.

Resource Constraints. As early as January 1989—before the construction of the Base Force—an economic slowdown and the growing federal deficit had led Office of Management and Budget (OMB) Director Richard Darman to press for substantial defense cuts in the FY 1990 budget request.[4] At the same time, the Joint Chiefs of Staff (JCS) sought a real 2 percent increase in defense budgets. The White House chose a middle path and froze spending for one year but planned a real 1.2 percent increase over the Future Years Defense Program (FYDP), with the expectation that the threat environment and resulting defense needs would be clearer in a year's time. Although actual cuts were avoided, desires for reductions to defense spending as the Base Force was being developed drew additional impetus from another development: the possibility that crippling spending cuts would automatically be triggered under the Gramm-Rudman antideficit law.[5] Under Gramm-Rudman, the fiscal 1991 deficit could not exceed $74 billion ($64 billion plus a $10 billion margin of error). If the deficit passed the $74 billion threshold, automatic sequestration (spending cuts) of discretionary spending would occur, half from defense and half from domestic spending. The final budget reconciliation bill passed in 1990 and—unanticipated by the Base Force plan—included approximately $184 billion in cuts from appropriations bills, with defense taking the

---

[3]These appear to have been related primarily to supporting the Pacific Command (PACOM) and exercises.

[4]For a discussion of these early budget pressures, see Snider, *Strategy, Forces and Budgets*, pp. 10–11. See Table A.1 in the appendix for data on federal deficits or surpluses from FY 1981–2000. The savings-and-loan crisis was one major cause of the increased deficit during this period. By early 1989, approximately one in six savings and loans was bankrupt. As a result, Congress in August 1989 passed a $50 billion bailout plan (P.L. 101-73), but those funds soon proved inadequate; by 1990, the administration was projecting that as much as $130 billion in taxpayer funds might be needed to cover thrift losses, and by January 1992, the Congressional Budget Office was estimating the budgetary cost at $200 billion. See Congressional Budget Office, *The Economic Effects of the Savings and Loan Crisis*, Washington, D.C., January 1992.

[5]The following discussion is based on "Budget Adopted After Long Battle; Five-Year Plan Promises $496 Billion in Deficit Reduction," *1990 CQ Almanac*, Washington, D.C.: Congressional Quarterly Press, 1991, pp. 111–166.

largest cut. In fact, defense provided *all* of the cuts in discretionary spending in the first three years, totaling $67.2 billion.

Strategy Under the Base Force

Strategic Precepts. The Base Force was developed under Chairman of the Joint Chiefs of Staff (CJCS) Colin Powell in parallel with the Bush administration's large-scale review of national security and defense strategy under National Security Review 12 (NSR 12).[6] The aim of the Base Force was to develop a new military strategy and force structure for the post–Cold War era while setting a floor for force reductions, in large part to hedge against the risks of a resurgent Soviet/Russian threat. The Base Force and the national security review were both predicated on the assumption of a 25 percent reduction in force structure and a 10 to 25 percent reduction in defense resources.[7]

As described by Chairman Powell, the Base Force was to be the minimum force needed to execute the new strategy, preserve U.S. leadership, protect U.S. interests, and meet enduring defense needs.[8] The overall nature of the reductions and the shape of the resulting force would be constrained by Powell's desire to avoid "breaking" the force and to secure the backing of the service chiefs. This led to a

---

[6]Under the Bush administration, a review of national defense strategy was conducted by the National Security Council (NSC) between January and June 1989. The defense review was conducted by two principal committees: one addressing future arms control negotiations, strategic forces, and targeting, chaired by Arnold Kanter of the NSC staff, and the other on defense policy, strategy, and nonstrategic forces, chaired by Under Secretary of Defense for Policy Paul Wolfowitz. See Snider, *Strategy, Forces and Budgets*, p. 19.

[7]In 1989, JCS planners anticipated a 25 percent reduction in defense budgets, while the office of the Secretary of Defense anticipated only a 10 percent decline. See Jaffe, *The Development of the Base Force, 1989–1992*, p. 9. The JCS planners were later proven correct.

[8]See Colin L. Powell, "The Base Force: A Total Force," presentation to the Senate Appropriations Committee, Subcommittee on Defense, September 25, 1991. According to press reports, in the spring of 1992 some consideration was given in the Defense Planning Guidance (DPG) to a strategy that would aim to block the emergence of new superpowers that might challenge the United States. As a result of the criticism that followed, the DPG was revised to remove this goal. See William Matthews, "Soviet Demise Leaves Pentagon Wondering Who Is the Foe," *Defense News*, February 24, 1992, and Patrick E. Tyler, "Pentagon Drops Goal of Blocking New Superpowers," *New York Times*, May 24, 1992, p. 1.

certain amount of "fair sharing" of budget and manpower reductions and mitigated against a more imaginative or revolutionary transformation of the force.

Defense Secretary Richard Cheney's review of past drawdowns in late 1989 and early 1990 animated a concern about avoiding the cautionary lessons of history. Secretary Cheney sought a very different outcome for the post–Cold War force than those that had characterized the years after World War II, Korea, and Vietnam, which had been handled in a rushed and somewhat haphazard fashion. These drawdowns had all adversely affected the force, most recently in the form of the so-called hollow forces of the post-Vietnam era. To accomplish his post–Cold War build-down in a manner that would ensure the health of the force, Secretary Cheney formed a strategic alliance with Chairman Powell that came at the price of recognizing the chairman's own constraints.

The Bush administration's new defense strategy was first announced by President Bush in his address to the Aspen Institute on August 2, 1990, the day Iraq invaded Kuwait.[9] In this address, President Bush announced the replacement of the Cold War strategy—deterrence of Soviet aggression and coercion against America and its allies across the conflict spectrum—with a new strategy based on regional threats. The president also detailed the implications of this change for U.S. military forces: a 25 percent reduction in active forces and a need to reshape those forces for the post–Cold War era.

Although the Base Force was presented in detail in congressional hearings throughout 1991, it was not until the 1992 National Military Strategy (NMS) that the numerous and complex linkages between national security strategy, national military strategy, and the Base Force's force structure were described in full detail.[10] The 1992 NMS identified four "foundations" for national military strategy:

---

[9]The ideas expressed in President Bush's August 2, 1990, speech at the Aspen Institute were given fuller exposition in the August 1991 and January 1993 releases of the National Security Strategy. The ramifications for military strategy also found expression in the other sources that were described earlier.

[10]Most of the core elements of the 1992 NMS can be found in Powell's earlier briefings, or in Joint Chiefs of Staff, *Joint Military Net Assessment* (JMNA), 1991 and 1992.

- **Strategic deterrence and defense.** The NMS identified as an enduring defense need strategic forces that preserved stable deterrence through a modernized offensive force structure and continued research on defenses.[11] Of the four strategic concepts, only this one can truly be considered to be a carryover from the Cold War.

- **Forward presence.** The concept of forward presence in key areas was inherited from the 1989 NMS but was given additional emphasis in the Base Force. This concept, which in effect replaced the earlier Cold War concept of forward defense, called for smaller permanent forces, together with periodic deployments, to demonstrate U.S. commitment to protecting its interests overseas.[12]

- **Crisis response.** The reductions in forward-deployed forces necessitated improvements in the capability of U.S.-based forces to respond to crises. The NMS also introduced the need for sufficient forces to deter a second conflict when preoccupied by a major regional contingency.[13]

- **Reconstitution.** Reconstitution was the capacity to rebuild forces if needed[14] and to "preserve a credible capability to forestall any potential adversary from competing militarily with the United States."[15]

These foundations were tied both to policymakers' assumptions about the sorts of military challenges that would need to be met in the future and to the force structure that would be necessary.

Assumptions About Future Operations. The Base Force was predicated on the assumption that the United States would not have to undertake any significant commitments of forward-deployed

---

[11] The Base Force study undertook a review of the strategic nuclear competition with the Soviet Union concurrently with its assessment of general-purpose forces. See Snider, *Strategy, Forces and Budgets*, p. 11.

[12] See Jaffe, *The Development of the Base Force, 1989–1992*, pp. 3–4.

[13] See Joint Chiefs of Staff, *1992 National Military Strategy*, p. 7.

[14] See White House, *National Security Strategy of the United States*, Washington, D.C., August 1991, p. 25.

[15] Joint Chiefs of Staff, 1992 *National Military Strategy*, p. 7.

forces—a belief suggesting that policymakers did not envision the sorts of long-duration contingency operations that would ultimately prevail during that decade.[16] Such an expectation would have been in keeping with the Cold War experience, in which crisis deployments and intervention operations were occasionally conducted, but in which large-scale, long-term commitments to contingency operations were generally avoided.[17]

The Spectrum of Threat. There appears to have been general acceptance among policymakers that the so-called spectrum of threat was an accurate characterization of the likely future distribution of threats and military operations that the United States was likely to face.[18] In the spectrum-of-threat construct, the probability of occurrence and level of violence are inversely related, while the consequence of failure is positively related to the level of violence. The result is that humanitarian and disaster relief operations and other peacetime operations are more probable but generally less consequential than lesser regional conflicts (LRCs); high-intensity conventional major regional conflicts (MRCs) are less likely but more consequential than LRCs; and MRCs are more probable but far less consequential than global thermonuclear war.

Closely related to the spectrum of threat are the assumptions the Base Force made regarding the need for forces prepared to meet demands across the entire spectrum of contingencies; in this, the Base Force generally took a fairly traditional approach. As Figure 2.2 shows, policymakers anticipated that during peacetime U.S. forces would be engaged in forward presence operations. In the event of a crisis, it was expected that forward presence forces would be drawn down and deployed to the crisis, and crisis response forces based in the United States would be moved forward. In the extreme case of global warfighting, not only would all available forces in the Base

---

[16]See Richard Cheney, testimony before the Senate Armed Services Committee, January 31, 1992, and U.S. General Accounting Office, *Force Structure: Issues Involving the Base Force*, p. 3.

[17]Obvious exceptions were the major regional wars of the Cold War (Korea and Vietnam) and, to a lesser extent, U.S. involvement in the Chinese civil war in 1945–1949.

[18]See, for example, the discussion of risk in Joint Chiefs of Staff, *Joint Military Net Assessment*, 1991, pp. 12-3 to 12-5.

The Base Force: From Global Containment to Regional Forward Presence   13

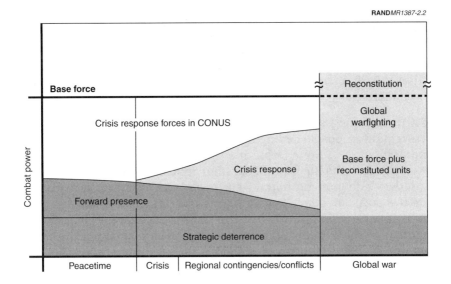

Figure 2.2—The Base Force and the Spectrum of Conflict (1992 JMNA)

Force be employed, but others would also be reconstituted to provide additional capability.

It is important to note that the multiple concurrent major theater war (MTW) construct that was to dominate defense planning for the remainder of the decade was not believed to be an intrinsic capability of the Base Force but was instead an afterthought. As late as February 1992, Chairman Powell testified that the 1997 force would be able to accommodate one MRC "with great difficulty" but that two concurrent Desert Storm and Korean campaigns would put the force "at the breaking point."[19] In fact, it appears that concurrent Persian Gulf and Korean contingencies were not included as illustrative planning scenarios (IPSs) until the FY 1994–1999 Defense Planning Guidance (DPG). Although the origins of the two-MTW standard were inauspicious, they would, with the BUR and QDR, come to constitute high canon for defense planning.

---

[19]See Colin Powell, testimony before the House Armed Services Committee, February 6, 1992.

The Role of Peacekeeping. With the reduction in tensions and ultimately the fall of the Soviet Union, it became much easier for the United Nations Security Council (UNSC) to authorize peacekeeping operations. One result was that the number of active U.N. operations greatly increased over their historical average of six or fewer operations per year, particularly after 1990.

The record suggests that the Bush administration supported U.N. involvement in peacekeeping operations and was willing to increase nonmilitary (i.e., State Department) funding for these operations.[20] Nevertheless, the Bush administration appeared to take an interested but decidedly noncommittal view toward U.S. military participation in the sorts of peace enforcement operations implied by U.N. Secretary General Boutros Boutros-Ghali's "agenda for peace."[21] In particular, the administration assiduously avoided making any broad commitments of U.S. forces to combat roles in future U.N. peace operations.

The View from the Air Force. On the whole, the Air Force chose to adapt quickly to the tumultuous changes in the strategic and budgetary environments and to shape the debates of the period.[22] In June 1990, it released a white paper, *Global Reach–Global Power*, that provided a review of more than 600 operations conducted by the Air Force since 1947 and that demonstrated the substantial contributions made by USAF combat, airlift, C^4ISR, and other capabilities across the entire spectrum of threat.

---

[20]The FY 1993 President's Budget request proposed an increase to the State Department's Conduct of Foreign Affairs account (153) funding for U.N. peacekeeping from $141 million in budget authority in 1989 to $438 million in FY 1993. Meanwhile support for peacekeeping under the International Security Assistance account (152) fell slightly, from $32 million in FY 1989 budget authority to $27 million in FY 1993. See Office of Management and Budget, *Budget of the United States Government, Fiscal Year 1993*, Washington, D.C., January 1992, p. 245. In 1993, $9 million was proposed for the U.N. force in Cyprus and $18 million for the Egypt/Israel/United States peacekeeping force in the Sinai Peninsula. Op. cit., p. 243. However, the administration was not entirely successful in ensuring that Congress actually funded the U.N.

[21]See Boutros Boutros-Ghali, *An Agenda for Peace: Preventive Diplomacy, Peacemaking and Peacekeeping*, New York: United Nations, June 17, 1992.

[22]See the charts providing a chronology of key events and Air Force actions in Donald B. Rice's foreword to U.S. Air Force, *Toward the Future: Global Reach–Global Power: U.S. Air Force White Papers, 1989–1992*, Washington, D.C., January 1993, pp. ii–iii.

The Air Force's role in U.S. national strategy was defined as sustaining deterrence; providing versatile combat forces; supplying rapid global mobility; controlling the high ground; and building U.S. influence through training, exercising, and participating in other activities with allied and other armed forces.[23] In the post–Cold War era, the Air Force could be expected to emphasize "global situational awareness; rapid, long-range power projection; the ability to deploy quickly and go the distance unconstrained by geography; and the range of lethal or peacetime actions to build U.S. influence abroad."[24]

The Air Force sought to adapt itself to better support the new regional strategy, which was predicated on assumptions that played to Air Force strengths: a reduction in forward-deployed forces and heavier reliance on rapid power projection and long-range strike operations from the United States. Indeed, the Base Force strengthened air power's role in future force packages, apparently out of the belief that U.S. forces could favor air power and shift away somewhat from heavy ground forces, particularly tanks.[25]

Building the Force

The Base Force assessment process benefited significantly from being embedded in the machinery of the Planning, Programming, and Budgeting System (PPBS).[26] This process involved multiple actors (J-5 and J-8 in the Joint Staff and various offices of the Office of the Secretary of Defense [OSD]) taking independent yet complementary

---

[23] See U.S. Air Force, *45 Years of Global Reach and Power: The United States Air Force and National Security: 1947–1992, A Historical Perspective*, Washington, D.C., 1992, pp. 2–4.

[24] See U.S. Air Force, *Toward the Future: Global Reach–Global Power*, p. i.

[25] See U.S. General Accounting Office, *Force Structure: Issues Involving the Base Force*, p. 29. Nevertheless, the Base Force was also predicated in part on the assumption implied in the Army's review of its warfighting doctrine, "Airland Battle," that future wars would involve significant clashes of armor against armor. Ibid.

[26] For a detailed discussion of how the Base Force was evaluated through the PPBS process, see Lewis, Roll, and Mayer, *Assessing the Structure and Mix of Future Active and Reserve Forces: Assessment of Policies and Practices for Implementing the Total Force Policy*, and Rostker et al., *Assessing the Structure and Mix of Future Active and Reserve Forces: Final Report to the Secretary of Defense*.

cross-cuts at the problem of defining a force structure for the post–Cold War world, with a focus on capabilities, risks, and costs. By all accounts, the PPBS process resulted in a thorough appraisal of the Base Force.[27]

From Concept to Force Structure. The size of the Base Force was determined principally by the need to protect and promote U.S. interests in regions vital to the United States and not, as the BUR and QDR would be, on the basis of its capability to fight multiple MRCs.[28] The "Base Force concept" comprised four force packages—one of which consisted of strategic forces and the other three of conventional forces—and four supporting capabilities.

Although the force structure details were to be worked out through arms reduction agreements, *strategic forces* were to continue underwriting nuclear deterrence through reliance on a smaller triad of land-based, sea-based, and air-breathing strategic nuclear offensive forces and strategic defenses. As a result of progress on strategic arms control with the Soviets and Russians, by August 1992 the shape of the planned force was as described in Table 2.1.

The principal aims of the Base Force's *conventional forces* were to achieve conventional deterrence and to promote stability and otherwise shape the global environment while preventing the emergence of power vacuums that could lead to instability and militarization by hostile countries.[29] Conventional forces were essentially to be built from the bottom up, based on regional interests, according to a "threat-based" review of anticipated and potential threats and an appraisal of the self-defense capabilities of U.S. friends and allies.[30]

---

[27]The 1992 JMNA was to put the Base Force through a slightly different set of scenario-based assessments.

[28]Although it was not designed on this basis, the Base Force would, however, be assessed by the Joint Chiefs of Staff in terms of its ability to fight one or more MRCs.

[29]See U.S. General Accounting Office, *Force Structure: Issues Involving the Base Force*, p. 22.

[30]One conclusion of this review seems to be that the United States would by 1997 retain larger forces than any other country except China while continuing to spend substantially more on defense capabilities than any other nation. See U.S. General Accounting Office, *Force Structure: Issues Involving the Base Force*, p. 22, and U.S.

The Base Force: From Global Containment to Regional Forward Presence   17

Table 2.1
Proposed Strategic Forces Package as of August 1992

Force	Number
Offensive[a]	
SSBNs	18
ICBMs:	550
Minuteman III	500
Peacekeeper	50
Bombers	180
Defensive	
Theater ballistic missile defense	—
Global Protection Against Limited Strikes (GPALS)	—
Air defense squadrons	10

SOURCES: Department of Defense, *1992 Joint Military Net Assessment*, Washington, D.C., August 1992, pp. 3-1 through 3-10, and Department of Defense, *Annual Report to the President and Congress*, Washington, D.C., Table C-1, 1991, 1992, and 1993.

[a]SSBN = strategic ballistic missile submarine; ICBM = intercontinental ballistic missile.

Three force packages comprised the core of the Base Force's conventional force structure:

- **Atlantic forces.** Atlantic forces were to meet threats and secure interests across the Atlantic, primarily in areas of vital interest to the United States: Europe, Southwest Asia, and the Middle East.[31] These forces were to be "heavy," were to be oriented toward projection and reinforcement, and were to have a significant reserve component. Atlantic forces consisted of forces for forward presence and those for contingency response.

- **Pacific forces.** The objective of Pacific forces was to protect and promote U.S. interests in East Asia and the Pacific. Pacific forces were to be "light" and predominantly maritime and were to include some Army and Air Force forward-deployed presence, some ability to reinforce from the United States, and less of a reserve component than the Atlantic forces.

General Accounting Office, *National Security: Perspectives on Worldwide Threats and Implications for U.S. Forces*, GAO/NSIAD-92-104, April 16, 1992.

[31]Along with the contingency forces, Atlantic forces would presumably be available for use in Africa and the Western hemisphere as well.

- **Contingency forces.** Contingency forces were to consist of light, mobile forces that were to be CONUS-based and "ready to go on a moment's notice."[32] These rapidly mobile, highly lethal forces were seen as likely to serve as the leading edge of forces being introduced for major regional contingencies and were to be less reliant on reserve components than the Atlantic and Pacific forces.

When aggregated, the Atlantic, Pacific, and contingency force packages resulted in the core of the Base Force (see Table 2.2). This force

Table 2.2

Proposed Base Force Conventional Force Packages as of August 1992

Force Package	Army Divisions	USAF TFWEs	USMC MEFs	Navy Carriers
Atlantic				
Europe	2	3.42	0	2
United States				
Active	3	1.33	1	4
Reserve	6	11.25	1	0
Cadre	2	0	0	0
Subtotal	13	16.00	2	6
Pacific				
Japan	0	1.25	1	1
South Korea	1	1.25	0	0
United States	1	1.00	0	5
Subtotal	2	3.50	1	6
Contingency				
United States	5	7.00	1	0
Total	20	26.5	4	12

SOURCES: Joint Chiefs of Staff, *Joint Military Net Assessment*, Washington, D.C., August 1992, pp. 3-1 through 3-10; U.S. General Accounting Office, *Force Structure: Issues Involving the Base Force*, Washington, D.C., GAO/NSIAD-93-65, January 1993, Table 2.1, p. 17; and John M. Collins, *National Military Strategy, the DoD Base Force, and U.S. Unified Command Plan: An Assessment*, Washington, D.C.: Congressional Research Service Report 92-493S, June 11, 1992, Figure 3, p. 25.

---

[32]See statement of General Colin Powell, Chairman, Joint Chiefs of Staff, in U.S. Senate, Committee on Armed Services, Department of Defense Authorization for Appropriations for Fiscal Years 1992 and 1993, February 21, 1991, p. 44.

comprised 20 Army divisions (12 active, 6 reserve, and 2 reserve cadre divisions), 26.5 USAF tactical fighter wing equivalents (TFWEs) (15.25 active, 11.25 reserve), four Marine Expeditionary Forces (MEFs), and 12 Navy carriers. These force packages entailed a reduction in major force elements and manpower ranging from 20 to 40 percent, depending on the service, component, and force element.

Supporting forces included *transportation* capabilities and prepositioning designed to provide the capabilities necessary to rapidly project and sustain U.S. power projection; *space* capabilities to provide early warning, surveillance, navigation, C^3, and other services; *reconstitution* capabilities to provide a broad foundation of industrialization, mobilization, and sustainment that could be rapidly activated; and *research and development* (R&D) capabilities to provide an ongoing and vital foundation for the technologies, applications, and systems of the future that would ensure that the United States retained its technological superiority.

USAF Force Structure Issues. The Air Force's principal aim throughout the Base Force was to preserve its modernization and acquisition programs.[33] Accordingly, early in the process of defining the Base Force, Air Force leaders accepted the fact that the Air Force's force structure would be reduced and therefore focused on shaping the ultimate force levels.[34] The Base Force also necessitated a reduction in active manpower for the Air Force to approximately 436,400 by FY 1997 (a 20.3 percent decline compared with FY 1990 levels) and a reduction in reserve end strength to some 200,500 (a 21.6 percent decline).[35]

---

[33] For example, the 1991 JMNA (Joint Chiefs of Staff, 1992, pp. 3-4 to 3-5) notes: "Generally, the Air Force continues to trade force structure for modernization, preserving a flexible and modern force capable of absorbing new systems in response to future needs." Lewis, Roll, and Mayer and Rostker et al. confirm the Air Force's willingness to trade force structure for modernization.

[34] At the time, for example, the Air Force, which was planning to create composite wings that would include fighter/ground attack aircraft and intelligence, surveillance, and reconnaissance (ISR) capabilities, argued that the total number of such wings should be 26.5 rather than the 24 TFWEs then being discussed.

[35] See Powell, "The Base Force: A Total Force," slide 14.

Resources

Base Force decisions on resources reflected a combination of constrained top lines and decisions to realign spending priorities.

DoD Top Line. Pressures to reduce the defense top line came from two main sources: the OMB and negotiations with Congress on deficit and spending caps. OMB Director Darman's efforts to reduce defense spending began as early as late January 1989, well before fissures began to emerge in the Soviet empire. Darman's initial efforts were only partially rewarded in the FY 1990 President's Budget submitted in April 1989, which rejected a JCS request for real increases of 2 percent and instead froze spending levels for one year but planned a modest 1.2 percent annual real increase in defense spending over the course of the defense program. The FY 1991 President's Budget, however, planned real reductions in defense spending of 2 percent a year. By the time of the June 1990 budget summit, the Base Force was offered as an illustrative example of a 25 percent smaller force that could provide savings of about 10 percent. It was not until the October 1990 budget summit, however, that deficit and spending targets that necessitated deeper defense budget cuts were agreed to. The spending plan presented with the Base Force reflected this agreement.

As described in the spending plan submitted with the Base Force, policymakers anticipated spending reductions that were generally in line with a 25 percent reduction in forces: a decline of about 22.4 percent in DoD budget authority by FY 1995, when Base Force force structure targets were to be achieved, and 25.3 percent by FY 1996.[36]

DoD Priorities. Over the FY 1991–1993 period, while the specifics of the defense budget and program changed somewhat, its basic priorities remained the same: to retain high-quality, ready, and capable forces even as force levels were reduced and to continue to make robust investments in longer-term capabilities that could ensure a qualitative edge even as spending on weapon modernization de-

---

[36]These spending plans will be discussed later under planning and execution of the Base Force.

clined.[37] The result was an effort to maintain high levels of research, development, test, and evaluation (RDT&E) even as procurement spending plummeted.

With regard to procurement, it is worth noting that the procurement spending cuts in the FY 1992 program that implemented the Base Force were in fact deeper than had originally been anticipated. As was described above, the Base Force was decided before the explicit discretionary budget limits were known; only after the October 1990 budget summit, when a defense top line was established, was it clearly understood that the Base Force plan and the budget did not match. Accordingly, procurement cuts were largely used to bring the budget back into line, since the force structure cuts and associated operating savings alone would have been inadequate to produce this result.

Table 2.3 describes the notional allocation of resources that was to underwrite this dual focus on readiness and long-term modernization, as conceived by the FY 1992 budget submitted in February 1991. As this table shows, longer-term investment accounts (RDT&E) were to increase as procurement spending declined. Meanwhile, high readiness levels were to be funded by holding the line on military pay and operations and maintenance (O&M) funding.

In the event of deeper spending reductions in the future, Base Force policymakers expressed a willingness to continue trading modernization and force structure. In September 1991, for example, Chairman Powell described the likely strategy that would be followed in the event that defense resources continued to decline beyond the levels necessary to sustain the Base Force:

> As the budget drops, operations and maintenance expenditures in combat units will be further reduced. Bases will close. Certain overseas commitments will be reduced. Procurement, research,

---

[37]For example, in President Bush's 1992 State of the Union address, the president announced that spending for strategic nuclear forces and weapon modernization could be reduced $43.8 billion for FY 1993–1997 compared to what the administration had proposed the preceding year.

Table 2.3

DoD Budget Authority by Title, FY 1990 and FY 1993

Account	1990 ($B)	1993 ($B)	Percent 1990	Percent 1993
Investment				
RDT&E	36.5	41.0	12.5	14.8
Procurement	81.4	66.7	27.8	24.0
Military construction	5.1	3.7	1.7	1.3
Operations and support				
Military pay	78.9	77.5	26.9	27.9
O&M	88.3	84.7	30.1	30.5
Family housing	3.1	3.6	1.1	1.3
Total	293.0	277.9	100.0	100.0

SOURCE: Colleen A. Nash, "Snapshots of the New Budgets," *Air Force Magazine*, April 1991, p. 65.

and development programs will be further curtailed. Operating stocks, the basis for combat sustainability, will be reduced. Next, elements of the base force will be reduced to cadre status. Finally, despite the risk to national security, components of the base force will be eliminated. But never, never, will a single combat unit be unable to perform its mission due to lack of personnel or equipment.[38]

Air Force Priorities. As described above, the Air Force had assessed its core competencies early on and had aligned its priorities to meet the needs of the new environment and strategy as well as to meet the spending targets. Maintaining and enhancing the key contributions of the Air Force dictated a careful balancing act involving short-term considerations of readiness on the one hand and longer-term investments in next-generation air and space capabilities on the other, described later in this chapter.

---

[38] See Powell, "Building the Base Force: National Security for the 1990s and Beyond," slide 22.

IMPLEMENTING THE BASE FORCE

Although planning for the Base Force extended to FY 1995–1997 (force structure targets were to be achieved by FY 1995, with some manpower and other targets achieved by FY 1997), the outcome of the presidential elections of 1992 meant that the Base Force was in fact implemented only over the course of two years—FY 1992 and 1993—and in the latter year by the Clinton administration. This section discusses the planning and execution of some of the key Base Force concepts through FY 1993.

Strategy

In a sense, the first test—and affirmation—of the new regional strategy was the Iraqi invasion of Kuwait, which was precisely the sort of large-scale, mechanized cross-border regional aggression that the new strategy aimed to deter and defeat. Among the key lessons drawn from the Gulf War was that overwhelming force coupled with the qualitative edge afforded by high technology—including stealthy F-117s, conventional cruise missiles, precision-guided munitions, the Airborne Warning and Control System (AWACS), and the Joint Surveillance Target Acquisition Radar System (JSTARS)—could yield campaign outcomes that not only were quick and decisive but also could minimize U.S. casualties. U.S. aerospace power was overwhelming in the war, leading to the quick achievement of air supremacy,[39] the destruction and/or suppression of Iraqi integrated air defense systems (IADSs), and the dismantling of both Iraqi fielded forces and Iraqi command and control and other strategic capabilities.

The post–Gulf War experience also validated policymakers' assumptions about one implication of the "spectrum of threat": that U.S. forces might be kept exceedingly busy with a host of smaller and generally less consequential military operations. A large number of small operations were conducted in 1991–1992, including noncom-

---

[39]U.S. air superiority capabilities led Iraqi aviators to decline to fly and to the defection of a number of Iraqi aircraft to Iran.

batant evacuation operations (NEOs)[40] and traditional humanitarian relief operations, generally in permissive environments.[41] However, several more complex operations were also undertaken in Southwest Asia, Bosnia, and Somalia that involved both humanitarian relief and the potential for combat.

Although apparently not recognized at the time, the post–Gulf War experience in Southwest Asia cast doubt on the premise that the United States had moved from a strategy of forward defense to one of forward presence: U.S. forces would continue to stand watch in the Gulf for the remainder of the decade, creating an unanticipated burden on the force to sustain that presence through rotational deployments.

Force Structure and Manpower

Force Structure. Table 2.4 describes the planned changes under the Base Force out to 1997, and Table 2.5 provides an overview of the force structure changes that were planned through 1993 at the time of the FY 1992 President's Budget request, submitted in the spring of 1991. Table 2.5 compares the initial plans with the actual changes that took place over the period.

Table 2.5 shows that with few exceptions, the 1993 levels tracked closely with the plans released in spring 1991;[42] with these exceptions, implementation of the Base Force reductions to force structure appears to have gone more or less as planned through FY 1993.

Manpower. The Base Force anticipated reductions of roughly 25 percent in active and reserve manpower by 1997. Unlike the major force elements just described, however, manpower reductions did

---

[40] Between June 1991 and September 1992, NEOs were conducted in the Philippines, Zaire, Haiti, Sierra Leone, and Tajikistan.

[41] For example, between April 1991 and December 1992, humanitarian assistance operations were conducted in Turkey, Bangladesh, the Philippines, Cuba, the Commonwealth of Independent States, Bosnia, Somalia, and Guam. Additionally, troops were sent to lend humanitarian assistance in the wake of the Los Angeles riots in May 1992, Hurricane Andrew in Florida in September 1992, and Typhoon Iniki in September 1992.

[42] The notable exceptions are substantial reductions in the Poseidon-Trident missiles and, more generally, more modest reductions in naval forces.

Table 2.4

Planned Base Force Changes to Force Structure and Manpower, FY 1990–1997

Service and Major Forces	FY 1990	FY 1997[a]	Change
Army			
Army divisions	28	20	−8
Active	18	12	−6
Reserve[b]	10	8	−2
Navy			
Aircraft carriers	15	12	−3
Active	13	11	−2
Reserve	2	1	−1
Battle force ships	546	451	−95
Air Force			
Tactical fighter wings	36	26	−10
Active	24	15	−9
Reserve	12	11	−1
Strategic bombers	268	180	−88
Manpower (thousands)			
Active military	2070	1626	−444
Reserve military[c]	1128	920	−208
Civilian	1073	904	−169
Total	4271	3450	−821

SOURCE: Congressional Budget Office (CBO).

[a] The CBO assumed that planned forces for FY 1997 were the same as those for FY 1995.

[b] The 1997 reserve Army divisions include two cadre divisions.

[c] The 1990 reserve military number does not include some 26,000 members of the selected reserve activated for Operation Desert Storm. They are included in the 1990 active manpower total.

not take place according to plan, with reductions to active forces exceeding those planned and reductions to reserve forces failing to reach planned levels. For example, whereas the FY 1992 budget request implementing the Base Force planned for a reduction of 13.2 percent in total active military personnel between FY 1990 and FY 1993, the actual reduction was 21.4 percent. Meanwhile, selected reserve personnel, which were anticipated to decline by 12.3 percent, fell only about 6.2 percent over the period.

Table 2.5

Base Force Planned vs. Actual Force Structure Changes, FY 1990–1993

Service and Major Forces	Actual 1990	Planned 1993	Actual 1993	Actual Percent Change 1990–1993
Strategic forces				
ICBMs/fleet ballistic missiles				
Peacekeeper	50	50	50	0
Minuteman	950	800	802	–15.6
Poseidon-Trident	608	496	408	–32.9
Strategic bombers (PAA)[a]	244	168	169	–30.7
General-purpose forces				
Land forces				
Active Army divisions	18	14	14	–28.6
Reserve Army divisions	10	8	8	–25.0
Active Marine divisions	3	3	3	0
Reserve Marine divisions	1	1	1	0
Naval forces (total)				
Total naval vessels	545	464	448	–21.7
Aircraft carriers (deployable)	13	13	13	0
Battleships	4	—	—	—
Other major surface combatants	199	144	133	–49.6
Nuclear attack submarines	93	90	89	–4.5
Amphibious assault ships	63	58	55	–14.5
Sealift fleet (nucleus fleet)	70	70	63	–11.1
Air forces				
Active Air Force TFWEs	24	16	16.1	–49.2
Reserve Air Force TFWEs	12	11	11.5	–4.0
Active Navy carrier air wings	13	11	11	–18.2
Reserve Navy carrier air wings	2	2	2	0
Active Marine air wings	3	3	3	0
Reserve Marine air wings	1	1	1	0
Air Force conventional B-52 squadrons	2	2	2	0
Intertheater airlift (PAA)	400	386	386	–3.6

SOURCES: Office of Management and Budget, *Budget of the United States Government, Fiscal Year 1992*, Washington, D.C., February 4, 1991, Table A-3, p. 185, and Department of Defense, *Annual Report to the President and Congress*, Washington, D.C., January 1993.

[a] PAA = Primary Aircraft Authorization.

Defense Reform and Infrastructure

To provide additional savings, the administration pursued defense reform and infrastructure reductions. The July 1989 Defense Management Review (DMR), for example, identified a large number of initiatives to reform the acquisition process, streamline and reduce regulations, and remove unnecessary management layers. Similarly, the 1989 BRAC identified 40 bases for closure, and the 1991 BRAC proposed an additional 50 bases for closure. Nevertheless, by late 1992–1993 concerns had arisen that not all of the anticipated $70 billion in savings from the DMR and BRAC rounds would be realized. As measured by the number of major and minor Air Force installations in the United States and abroad,[43] USAF infrastructure fell only slightly in comparison to force structure reductions.[44]

Modernization

As described above, much of the savings from the Base Force were to come from reductions to modernization and, more specifically, from reductions to procurement. Plans for FY 1992 included the cancellation of more than 100 weapon system programs, with total savings estimated at $81.6 billion over the FY 1992–1997 defense program. Notwithstanding Air Force leaders' hopes to trade force structure for modernization and acquisition,[45] a number of high-priority Air Force modernization programs were reduced or terminated during the course of—or as a result of—the Base Force, including the B-2 (reduced from 132 to 75 aircraft, and subsequently to 20 aircraft) and the C-17 (reduced from 210 to 120 aircraft).

---

[43] Among the major Air Force base closures in the 1991 BRAC round, for example, were Bergstrom Air Force Base, TX (active component only); Carswell Air Force Base, TX; Castle Air Force Base, CA; Eaker Air Force Base, AR; England Air Force Base, LA; Grissom Air Force Base, IN; Loring Air Force Base, ME; Lowry Air Force Base, CO; Myrtle Beach Air Force Base, SC; Rickenbacker Air Force Base, OH; Williams Air Force Base, AZ; and Wurtsmith Air Force Base, MI. Data are from Assistant Secretary of the Air Force (Financial Management and Comptroller), *United States Air Force Statistical Digest*, Washington, D.C., various years. Although the number of installations is a crude measure, the lag in reductions to infrastructure is borne out by other data and analyses.

[44] From 1990 to 1993, the Air Force's infrastructure spending fell from $40 billion to $33 billion in FY 1999 billions of dollars. Nevertheless, infrastructure spending as a percentage of total USAF total obligational authority (TOA) climbed, from 42 to 44 percent. See U.S. Air Force, *Air Force Strategic Plan, Vol. 2: Performance Plan Annex; Performance Measure Details*, Washington, D.C., February 1999, Table 2.B.3, "Total Infrastructure Spending."

[45] For example, the Air Force's planned investment program averaged 47.4 percent across the FY 1992–1997 Defense Program.

Nevertheless, efforts to improve the conventional capabilities of long-range bombers and to expand the capabilities of precision-guided munitions were begun as a result of the 1992 Bomber Roadmap and other initiatives. In a similar fashion, the 1992 Mobility Requirements Study (MRS) identified serious deficiencies in strategic mobility capabilities and advocated enhancements to airlift, sealift, and prepositioning programs.[46]

The View from the Air Force. For the Air Force, some of the key force structure reductions over the FY 1990–1993 period (e.g., combat aircraft) appear to have been executed as planned, although infrastructure reductions lagged, providing no additional savings that could be used to preserve modernization. Reductions in the USAF's force structure during this period varied greatly by component and element of force structure (see Figure 2.3), with active-component TFWs falling by roughly one-third but reserve-component personnel falling by only about 4 percent. By contrast, the number of USAF installations fell by less than 8 percent.

Resources

The most significant change in defense resources during the Base Force period was the declining defense top line, which resulted from the administration's efforts to control the ballooning federal deficit by using defense as the principal bill payer.

DoD Budgets: DoD-Wide Top Line. Although it was not until the October 1990 budget summit and the final negotiations on the Budget Enforcement Act that discretionary defense spending caps were finalized, Base Force policymakers anticipated reduced annual defense budgets. These reductions accelerated with each succeeding year's budget, however, subsequently forcing policymakers to make deeper cuts in modernization accounts than had originally been anticipated.

As early as January 1989, OMB Director Darman had begun pressing for substantial reductions to defense spending. As shown in the

---

[46]Put another way, many of the so-called force enhancements that would later be described in the BUR already appear to have been under way with the Base Force.

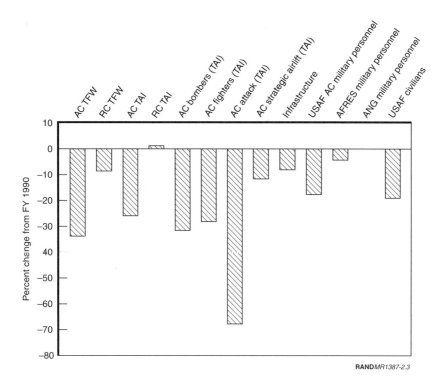

Figure 2.3—USAF Force Structure and Manpower Reductions, FY 1990–1993

"pitchfork" chart in Figure 2.4,[47] however, the FY 1990 President's Budget submitted in April 1989 envisioned a modest real annual increase of 1.2 percent. These planned increases were short-lived, and with the subsequent (FY 1991–1993) budget plans, reductions were programmed: Whereas the April 1989 plan for defense spending in FY 1990 had anticipated a 1.2 percent real increase, in subsequent years real declines were planned in the amount of 2 percent (the FY 1991 plan), 3 percent (FY 1992), and 4 percent (FY 1993). By January

---

[47]This chart, used often by the Bush defense department in early 1991, came to be called the "pitchfork" chart.

30 Defense Planning in a Decade of Change

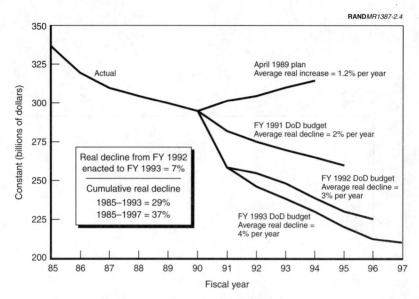

SOURCE: Joint Chiefs of Staff, *Joint Military Net Assessment*, Washington, D.C., August 1992.

Figure 2.4—"Pitchfork" Chart, Circa 1992

1993, the outgoing administration's budget anticipated a 21.4 percent decline from FY 1990 levels by the end of the fiscal year, compared with the 16.1 percent decline that had been anticipated in the initial (FY 1992) Base Force budget.

Figure 2.5 presents "FYDP tracks" that compare successive DoD long-term spending plans (the solid lines) with actual spending.[48] As Figure 2.5 shows, the administration's spending plans continued to fall over these years.[49] In FY 1992 and FY 1993 (the years in which

---

[48] The chart can thus be used to compare how well spending plans anticipated actual spending. In the chart, the solid lines reflect successive long-range DoD budget plans—the lines fall over time, reflecting reductions in each year's long-range budget plan. The small circles represent actual budget authority in current (then-year) billions of dollars.

[49] A total of $43.1 billion in budget authority and $27.1 billion in outlays were cut between the five-year plans submitted for FY 1992 and FY 1993. See Congressional

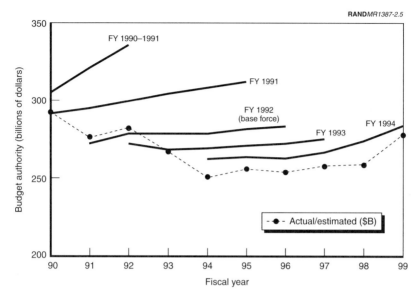

Figure 2.5—Long-Term DoD Budget Plans, FY 1990–1991 Through FY 1994

the Base Force was implemented), for example, actual spending closely approximated planned spending. In contrast, actual spending for 1994–1999 fell well below that planned by the Bush administration—a result of further defense budget reductions by the Clinton administration.[50]

There were also shifts in the allocations of DoD long-term budget authority. Operations and support (O&S) accounts, for example, grew from 58.6 to 62.6 percent of planned budget authority, while investment accounts declined from 41.0 to 38.5 percent. In percentage terms, the biggest loser was procurement, which fell from 26.2 to 21.7 percent of total budget authority, while the biggest winners were O&M and RDT&E. The cumulative result of these shifts is shown in Table 2.6.

Budget Office, *An Analysis of the President's Budgetary Proposals for Fiscal Year 1993*, Washington, D.C., March 1992.

[50]These will be discussed in the next chapter. The FYDP track for the FY 1994 plan is the outgoing Bush administration's long-range spending plan for FY 1994–1999, which would have been submitted had President Bush won the 1992 presidential election.

Table 2.6

Bush Administration Long-Term Defense Budget Plans, FY 1991–1994 (percentage of total budget authority)

Department of Defense—Military	1991–1995 Program (%)	1992–1996 Program (%)	1993–1997 Program (%)	1994–1999 Program (%)
Operations and support	58.6	59.4	60.8	62.6
Military personnel	26.9	27.5	27.4	27.7
Operations and maintenance	30.6	30.6	32.0	33.4
Family housing	1.2	1.3	1.4	1.5
Investment	41.0	40.7	38.7	38.5
Procurement	26.2	24.9	22.2	21.7
RDT&E	12.9	13.9	14.0	15.2
Military construction	1.9	2.0	2.5	1.6

SOURCES: Office of Management and Budget, *Budget of the United States Government*, Washington, D.C., various years, and Stephen Daggett, *A Comparison of Clinton Administration and Bush Administration Long-Term Defense Budget Plans for FY 1994–99*, Washington, D.C.: Congressional Research Service Report 95-20F, December 20, 1994.

NOTE: Major appropriation titles only. Percentages may not total 100 owing to rounding and to the exclusion of revolving and trust funds, offsetting receipts, and allowances.

Over the FY 1990–1993 period, the composition of the FYDP changed, with O&S's share rising and investment's share declining.

By 1993, O&S spending had fallen 11 percent below the FY 1990 levels, while investment accounts had fallen nearly 28 percent over the same period.

The View from the Air Force. As Tables 2.7 and 2.8 show, Air Force spending plans were also reduced over time (Table 2.7)[51]—and despite its desire to preserve modernization, the Air Force also saw a modest shift in the composition of Air Force spending from investment to O&S accounts (Table 2.8).

Within the investment accounts, procurement declined from 31 percent of total budget authority to 27 percent of USAF budget

---

[51] As shown, the decline from FY 1991–1992 seems largely to have been a postponement of reductions planned for FY 1991—an artifact of unexpectedly high spending in 1991 associated with the Gulf War.

Table 2.7
Planned vs. Actual USAF Spending (BA in billions of dollars)

	1990	1991	1992	1993
Budget Request ($B)				
FY 1990–1991	100	107		
FY 1991	93	95		
FY 1992		83	86	91
FY 1993			83	84
Actual Spending ($B)	93	91	82	79
Difference (planned – actual)	0.1	–8.6	0.2	4.7

SOURCE: DoD Comptroller, *National Defense Budget Estimates*, Washington, D.C., various years.

Table 2.8
Air Force Investment and O&S Spending, FY 1990–1993

Fiscal Year	FY 2000 Investment ($B)	FY 2000 O&S ($B)	Percent Investment	Percent O&S
1990	54	63	46	54
1991	43	64	40	59
1992	42	55	44	57
1993	40	51	44	56

SOURCE: DoD Comptroller, *National Defense Budget Estimates, FY 2000/2001*, Washington, D.C., 1999.

authority, while RDT&E increased slightly, from 14 to 16 percent. Meanwhile, military personnel increased from 26 to 27 percent and O&M from 27 to 28 percent over FY 1990–1993.[52]

---

[52] For a discussion of potential reductions to Air Force modernization accounts for the FY 1992–1993 budgets, see U.S. General Accounting Office, *1992 Air Force Budget: Potential Reductions to Aircraft Procurement Programs*, Washington, D.C., GAO/NSIAD-91-285BR, September 1991; U.S. General Accounting Office, *Air Force Budget: Potential Reductions to Fiscal Year 1993 Air Force Procurement Budget*, Washington, D.C., GAO/NSIAD-92-331BR, September 1992; and U.S. General Accounting Office, *1993 Air Force Budget: Potential Reductions to Research, Development, Test, and Evaluation Programs*, GAO/NSIAD-92-319BR, September 1992.

Challenges over the Horizon

Base Force policymakers recognized the challenges of attempting to balance strategy, forces, and resources in planning and executing the Base Force program and budget.[53] In fact, they appear to have anticipated a number of important program execution risks in the out years. For example, a December 1991 study by the Congressional Budget Office concluded that the size of the Base Force could probably be maintained through 1997 with the funding that the administration was projecting at the time, although some delays in programs for research, modernization, or other activities were seen as likely.[54] After 1997, however, substantial increases in spending—from $20 billion to as much as $65 billion by the middle of the next decade—could be needed to carry out the planned Base Force modernization.

According to the General Accounting Office, the principal challenges the Base Force faced were continued congressional pressure to further reduce defense budgets and the possibility that these cuts would bring defense resources below the levels that were necessary to sustain the force as planned.[55] By December 1992, the list of challenges facing DoD planners had grown. A "significant mismatch" between the $1.4 trillion FY 1993–1997 defense spending plan and budget realities had emerged, possibly necessitating additional program reductions of more than $150 billion.[56]

The Procurement "Bow Wave." A review by a Defense Science Board Task Force of the FY 1994–1999 FYDP suggested a number of potential funding shortfalls in the defense program.[57] The task force pro-

---

[53] As the 1991 JMNA put it: "Our assessment of the emerging world order suggests that meeting the demands of our global military objectives with fiscally constrained forces based largely within CONUS will continue to be an enormous challenge." See Joint Chiefs of Staff, *Joint Military Net Assessment*, 1991, pp. 2–3.

[54] See Congressional Budget Office, *Fiscal Implications of the Administration's Proposed Base Force*, Washington, D.C., December 1991.

[55] See U.S. General Accounting Office, *Defense Planning and Budgeting: Effect of Rapid Changes in National Security Environment*, Washington, D.C., GAO/NSIAD-91-56, February 1991, p. 1.

[56] See U.S. General Accounting Office, *National Security Issues*, Washington, D.C., GAO/OCG-93-9TR, December 1992, pp. 5–6.

[57] The task force was asked to assess (1) savings from the Defense Management Review Decisions (DMRDs); (2) development and acquisition costs for the weapons,

jected a shortfall in acquisition costs of $3 billion to $5 billion, although it had received information suggesting that acquisition shortfalls in the FY 1994–1999 Base Force program could be as high as $46.4 billion. Additionally, the task force predicted that the DoD faced a procurement "bow wave" of approximately $5 billion a year by the early 2000s that would in all likelihood make the planned theater air and Navy shipbuilding programs unaffordable after FY 1999. (See Figure 2.6 for a portrayal of the bow wave in the acquisition program as estimated in the BUR.)

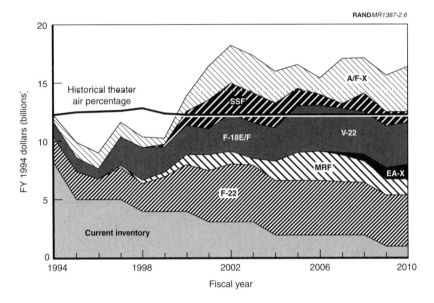

Figure 2.6—The Bow Wave in the Base Force Theater Air Program (BUR)[58]

---

sensors, and other major systems then in development, including any potential procurement bow wave; (3) O&M funding levels to support the planned force structure and projected personnel levels; (4) environmental cleanup and compliance costs; and (5) defense health care costs. See Defense Science Board, *Task Force on the Fiscal Years 1994–99 Future Years Defense Program (FYDP)*, reports of May 3, 1993, and June 29, 1993.

[58]Labeling in this figure is as it appeared in the BUR.

36 Defense Planning in a Decade of Change

The issue was never resolved by the Bush administration, and the bow wave in the theater air program would later be a subject for assessment by the BUR.

ASSESSMENT

The Bush Defense Department continued to promote the Base Force until the end of the administration, and there are few indications that, once defined, any alternatives were seriously considered. There is evidence, however, that the JCS recognized that changes in regional threats and balances of forces—not to mention defense budgets—had continued apace since the Base Force study was completed and that these changes might enable—or require—a new, even smaller Base Force.[59] Indeed, public hints of a new, smaller Base Force can be found in 1992–1993.[60] The August 1992 Joint Military Net Assessment (JMNA), for example, suggested that the Base Force was "designed to provide us with the capabilities needed to deal with an uncertain future; the Base Force is dynamic and can be reshaped in response to further changes in the strategic environment." Chairman Powell subsequently signaled further flexibility on the size of the Base Force: "Our Base Force is dynamic. There is nothing sacrosanct about its number of tanks, ships or missiles, its structure or its manpower."[61]

Capability to Execute the Strategy

The 1992 JMNA reported that on balance, the Base Force of 1999—if funded and carried out in accordance with the defense program—

---

[59] Among the changes in threat that were described in 1992, for example, were continued declines in the Soviet/former Soviet Union nuclear and conventional threats to Europe and the dismantling of the Iraqi war machine in the Gulf War.

[60] In testimony before the Senate Armed Services Committee on January 31, 1992, Chairman Powell said that when the U.S. military reached the Base Force targets in FY 1995, the country should debate whether it was the right force level. And in its comments on the U.S. General Accounting Office's report on the Base Force, the DoD characterized the Base Force as "dynamic [and] able to be reshaped (either upward or downward) if strategic developments warrant it." See U.S. General Accounting Office, *Force Structure: Issues Involving the Base Force*, p. 19.

[61] See Colin L. Powell, "U.S. Forces: Challenges Ahead," *Foreign Affairs*, Vol. 71, No. 5, Winter 1992–1993, pp. 32–45.

would be better capable of dealing with the uncertain post–Cold War era than the force in being at that time. The summary judgment of the 1992 JMNA was that the FY 1993 President's Budget request and defense program provided the U.S. Armed Forces with the minimum capability to accomplish national security objectives with low to moderate risk, which compared favorably with a higher level of risk during the Cold War. This capability was judged to be at increasing risk, however, as a result of key shortcomings stemming from declining investments in the industrial base, technology, and R&D and by prospects of even further cuts in force structure and capabilities.

In terms of the force's capability to respond to one or more MRCs, the 1992 JMNA determined that the Base Force was capable of resolving only one MRC at a time both quickly and with low risk; the risk to U.S. objectives in either individual MRC was judged to be moderate, but there was little margin for unfavorable circumstances.[62] In the event of two crises occurring closely together, it was judged that policymakers and commanders would have to employ economy of force and sequential operations and would need to make strategic choices regarding the apportionment of forces.[63]

The 1992 JMNA also rendered judgments on each of the Base Force's core capabilities:

- **Strategic deterrence and defense.**[64] U.S. offensive strategic forces were judged to provide sufficient capability for deterrence, but strategic defensive forces were seen to have only marginal capability, contributing primarily to early warning of strategic attack.

- **Forward presence.**[65] Reductions in basing and access rights were seen as a cause for concern in light of the judgment that

---

[62]The reader will recall that in testimony, Chairman Powell had indicated that two concurrent MRCs in Korea and Southwest Asia would put the force "at the breaking point." See Powell, testimony before the House Armed Services Committee, February 6, 1992.

[63]See Joint Chiefs of Staff, *Joint Military Net Assessment*, 1992, pp. 9-6 through 9-8 and 9-11 through 9-12.

[64]Op. cit., pp. 12-3 through 12-4.

[65]Op. cit., pp. 12-4 through 12-5.

forward presence and the ability to project power rapidly would become increasingly important. The capabilities for other operations, such as humanitarian, civic action, disaster relief, and counterdrug and counterterrorist operations, were judged to be less problematic.

- **Crisis response.**[66] Overall crisis response capabilities were deemed adequate throughout the assessment period and were expected to improve as specific deficiencies in mobility and force capabilities were eliminated.

- **Reconstitution.**[67] The capability for reconstitution was seen to require monitoring to ensure that key industrial base and other capabilities were not lost.

Readiness

In general, during the implementation of the Base Force, readiness indicators of all components seem to have improved. However, the ability to sustain high readiness rates was taxed by the increasing complexity of threats and missions and by the associated training requirements—an outcome that would likely have been more apparent had the Base Force continued into the late 1990s. The impact on reserve-component readiness could have been especially significant because of limited training time.

Modernization

As noted above, the Defense Science Board Task Force identified the potential for a number of funding shortfalls in the defense program. Particularly troubling in terms of modernization was the task force's prediction that the DoD would face a procurement bow wave of approximately $5 billion a year by the early 2000s.

---

[66] Op. cit., pp. 12-5 through 12-6.
[67] Op. cit., p. 12-6.

SECTION CONCLUSIONS

Many of the strategic assumptions underlying the Base Force would, with only modest adjustment, remain salient through the rest of the decade. Among the most important of these were the need for forces tailored to a post-Soviet, post–Cold War world and the focus on a regionally based strategy that emphasized deterrence, forward presence, and crisis response.

The Base Force also made significant progress toward its force structure goals. By FY 1993, the Air Force was less than one wing away from achieving the Base Force target of 15.25 TFWs, for example, and had already achieved its 11.25-wing target for reserve-component forces.

That said, one of the Base Force's key premises—that the post–Cold War world would not be occasioned by large-scale, long-duration contingency operations—was cast in doubt by the post–Gulf War stationing of Air Force tactical fighter and other aircraft in Southwest Asia: a commitment that, despite predictions to the contrary, would remain through the end of the decade.

In addition, defense resources continued to tumble even after the Base Force was defined, leading to a widening gap between strategy, forces, and resources and setting the stage for a number of hard choices that would need to be faced in the out years, with modernization and readiness of the force being the main ones.

While it cannot be known how Base Force policymakers might have addressed these issues in FY 1993–1994, the fact that they were never satisfactorily addressed and resolved—either by Bush administration or by Clinton administration policymakers—meant that they not only would remain for much of the rest of the decade but would ultimately exacerbate an emerging gap between strategy, forces, and resources.

Chapter Three
THE BOTTOM-UP REVIEW: REDEFINING POST–COLD WAR STRATEGY AND FORCES

The 1993 *Report on the Bottom-Up Review* was the second major force structure review of the decade.[1] The aim of the BUR was to provide "a comprehensive review of the nation's defense strategy, force structure, modernization, infrastructure, and foundations."[2] The BUR's force structure reductions were to accelerate and surpass those planned in the Base Force, leading to a total reduction in forces

---

[1] The BUR is documented in Defense Secretary Les Aspin's and Chairman Powell's briefing slides in "Bottom-Up Review," Washington, D.C., September 1, 1993; in the DoD's transcript of that briefing in Department of Defense, Pentagon Operations Directorate, "Bottom-Up Review Briefing by SECDEF and CJCS," Secretary of Defense Message P020023Z SEP 93, September 2, 1993; in the preliminary release of the BUR's results on the same date in Les Aspin, *The Bottom-Up Review: Forces for a New Era*, and Les Aspin, *Force Structure Excerpts: Bottom-Up Review*, Washington, D.C., September 1993, and in the final report, contained in Les Aspin, *Report on the Bottom-Up Review*, Washington, D.C., October 1993. A detailed discussion of the BUR's assessment process can be found in the testimony of Edward L. Warner III, Assistant Secretary of Defense for Strategy, Requirements, and Resources, before the House Armed Services Committee, February 2, 1994, and that of Rear Admiral Francis W. Lacroix, Deputy Director for Force Structure and Resources Division (J-8), Joint Staff, before the House Armed Services Committee, March 1, 1994. For an analysis and critique of the key assumptions used in the BUR and in the Nimble Dancer exercises that tested the BUR, see U.S. General Accounting Office, *Bottom-Up Review: Analysis of Key DoD Assumptions*, Washington, D.C., GAO/NSIAD-95-56, January 1995, and U.S. General Accounting Office, *Bottom-Up Review: Analysis of DoD War Game to Test Key Assumptions*, Washington, D.C., GAO/NSIAD-96-170, June 1996. An excellent history of the BUR can be found in Mark Gunzinger, "Beyond the Bottom-Up Review," in *Essays on Strategy XIV*, Washington, D.C.: Institute for National Security Studies, National Defense University, 1996, available at http://www.ndu.edu/inss (accessed September 2000).

[2] Aspin, *Report on the Bottom-Up Review*, p. iii.

from FY 1990 of about one-third—well beyond the Base Force's planned 25 percent reduction, most of which had already been achieved by the end of FY 1993. Budgets would also fall beyond planned Base Force levels as a result of the BUR.

The BUR redefined the meaning of engagement in an important way, giving increased rhetorical and policy importance to U.S. participation in multilateral peace and humanitarian operations while setting the stage for an increased operational tempo and rate of deployment even as force reductions continued. This chapter provides an assessment of the interplay between strategy, force structure, and resources in the development, design, and implementation of the BUR.

BUILDING THE BUR FORCE

Background

The World Situation. With the successful end of the Cold War and the dismantling of the Iraqi military in the Gulf War, together with continued progress on nuclear arms control with the states of the former Soviet Union, Clinton administration policymakers had some reason for optimism about the post–Cold War world. Upon their arrival in office, however, they found an unsettled environment filled with the sorts of challenges—in the former Soviet Union, Southwest Asia, Somalia, Bosnia, and elsewhere—that would occupy them for the remainder of the decade.

The U.S. military response to these ongoing challenges led to an increased commitment of Air Force aircraft to contingency operations. From January 1993, when the Clinton administration entered office, until September–October 1993, when the *Report on the Bottom-Up Review* was released, the average number of Air Force aircraft involved in contingency operations rose from roughly 175 to some 225 aircraft (see Figure 3.1). This increase reflected the early stages of executing the administration's activist conception of engagement, which was to be underwritten in part through the routine use of military forces in a wide range of forward presence operations.

Resource Constraints. During the 1992 presidential campaign, candidate Bill Clinton had argued that changes in the threat environ-

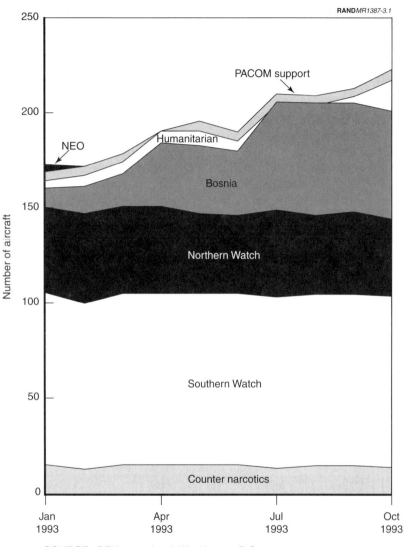

Figure 3.1—USAF Aircraft in Contingency Operations, 1/93–10/93

ment and the nation's poor economic circumstances[3] made possible a cut of approximately $60 billion in defense spending, or $10 billion a year over the FYDP.[4] Such a cut was consistent with what Clinton described as his first foreign policy priority for ensuring the United States' ability to lead: the restoration of America's economic vitality.

By the time of the FY 1994 budget submission in February 1993, the administration was planning force structure reductions to meet defense savings goals of $76 billion over FY 1994–1997 and $112 billion over FY 1994–1998.[5] As Table 3.1 shows, the cuts envisaged by the BUR were only slightly smaller than those documented six months earlier in the President's Budget. Put another way, the cuts to the defense top line planned in the FY 1994 budget were, within a few billion dollars in any given year, binding on the BUR.

Table 3.1

Evolution of Future Years Defense Programs in 1993
(BA in billions of dollars)

	1994	1995	1996	1997	1998	1999
1/93 Bush baseline[a]		257	261	264	270	273
Clinton administration:						
2/93 planned savings[b]	−7	−12	−20	−37	−36	—
FY 1994 plan[c]	251	248	240	233	241	
BUR plan[a]		249	242	236	244	250

[a]As recalculated and reported in the BUR.

[b]Defense discretionary spending changes as reported in Office of Management and Budget, *A Vision of Change for America*, Washington, D.C., February 17, 1993, Table 3-1, p. 22.

[c]As submitted by the Clinton administration in April 1993 and reported in DoD Comptroller, *National Defense Budget Estimates, FY 1994*, Washington, D.C., May 1993, Table 1-2.

---

[3]By June 1992, the federal deficit was estimated at $425 billion.

[4]See Bill Clinton, "A New Covenant for American Security," speech given at Georgetown University, December 12, 1991, and "Remarks of Governor Bill Clinton," Los Angeles World Affairs Council, August 13, 1992.

[5]See Office of Management and Budget, *A Vision of Change for America*, Washington, D.C., February 17, 1993, Table 3-1, p. 22.

As a result of these preexisting budgetary constraints, the strategy, force structure, modernization, and other initiatives described in the BUR were to be driven as much by the availability of resources as by the threats and opportunities in the emerging international environment documented by the BUR.

Strategy Under the BUR[6]

The national security strategy and force structure BUR was conducted in order to "develop guidelines for reducing and restructuring the U.S. defense posture in the context of a revised U.S. military strategy." This review was to be completed in time to publish the new DPG in July 1993, which would then be used by the services in their revisions to the FY 1994–1999 budget submissions. An OSD draft input to the administration's new national security strategy was completed on April 21,[7] while the assessment process seems to have begun in earnest in April and May 1993.[8]

The BUR provided the first coherent statement of the Clinton administration's strategy of "engagement, prevention, and partnership," which was to serve as a bridge to its first national security strategy statement the next year. In delineating its strategy, the BUR made clear that the administration did not feel bound by many elements of the national security and military strategies it had nominally inherited from the Bush administration. The BUR began by observing that "the Cold War is behind us. The Soviet Union is no longer. The threat that drove our defense decisionmaking for four and a half decades—that determined our strategy and tactics, our doctrine, the size and shape of our forces, the design of our weapons, and the size of our defense budgets—is gone."[9]

---

[6]This subsection draws heavily from Gunzinger, "Beyond the Bottom-Up Review."

[7]See Frank G. Wisner and Admiral David E. Jeremiah, "Toward a National Security Strategy for the 1990s," Washington, D.C.: Office of the Under Secretary of Defense for Policy, April 21, 1993, cited in Gunzinger, "Beyond the Bottom-Up Review."

[8]See Warner, testimony before the House Armed Services Committee, February 2, 1994, and Lacroix, testimony before the House Armed Services Committee, March 1, 1994.

[9]See Aspin, *Report on the Bottom-Up Review*, p. 1.

Having firmly placed the Cold War in the past, the BUR identified four principal "new dangers" facing the United States: the proliferation of weapons of mass destruction (WMD); regional dangers resulting both from large-scale aggression and from ethnic, religious, and other forms of conflict; threats to democracy and reform in the former Soviet Union and elsewhere; and economic instability resulting from the failure to build a strong and growing U.S. economy. The BUR saw the U.S. Armed Forces as central to combating the first two of these threats but also believed that the military could play a significant role in meeting the last two.

Assumptions About Future Operations. The BUR appears to have made three principal assumptions about future military operations. First, it assumed that the U.S. military would be very busy with peacetime operations in the post–Cold War world.[10] Second, it posited that U.S. forces would be engaged in operations in peacetime across the entire spectrum of conflict. Third, it assumed that peacetime demands could be managed so as to minimize impacts on the ability to conduct warfighting operations.

The BUR's planning strategy assumed that U.S. forces would need to be able to accomplish four major sets of objectives abroad:

- to defeat aggressors in MRCs;
- to maintain overseas presence to deter conflicts and provide regional stability;
- to conduct smaller-scale intervention operations such as peace enforcement, peacekeeping, humanitarian assistance, and disaster relief to further U.S. interests and objectives; and
- to deter attacks with WMD against U.S. territory, U.S. forces, or the territory and forces of U.S. allies.

These four objectives would be critical to the development of the BUR force structure, discussed next.

---

[10]Op. cit., Figure 6, "Conflict Dynamics," p. 27.

Building the Force

In order to address the four strategic objectives described above, the BUR recommended that forces be based on force building blocks—i.e., on distinct force packages for each objective:

- **Major regional contingencies.** The MRC building block was sized to fight a major regional conflict against a fairly substantial regional threat capable of launching an armor-heavy combined arms offensive against the outnumbered forces of a neighboring state.[11] The operational concept for the campaign was to undertake four phases of operations: (1) halt the invasion; (2) build up U.S. combat power in the theater while reducing that of the enemy; (3) decisively defeat the enemy; and (4) provide for postwar stability. The MRC force package consisted of 4 to 5 Army divisions; 4 to 5 Marine Expeditionary Brigades (MEBs); 10 Air Force fighter wings; 100 Air Force heavy bombers; 4 to 5 Navy aircraft carrier battle groups; and special operations forces.

- **Peace enforcement and intervention operations.** The second building block was oriented toward providing for a range of lower-intensity operations, from multilateral peace enforcement to unilateral intervention operations.[12] This building block was to be capable of forcing entry into, seizing, and holding key facilities; controlling troop and supply movements; establishing safe havens; securing protected zones from internal threats such as snipers, terrorist attacks, or sabotage; and preparing the environment for relief by peacekeeping units or civilian administrative authorities. The force package developed for this building block consisted of a total of 50,000 combat and support personnel.

---

[11]The canonical threat force for a single MRC consisted of 400,000 to 750,000 total personnel under arms; 2000 to 4000 tanks; 3000 to 5000 armored fighting vehicles; 2000 to 3000 artillery pieces; 500 to 1000 combat aircraft; 100 to 200 naval vessels; and 100 to 1000 Scud-class ballistic missiles, some possibly with nuclear, chemical, or biological warheads. Examples of such threats were Iraq (either prior to the Gulf War or following a posited rebuilding of its forces) and North Korea.

[12]Les Aspin, then chairman of the House Armed Services Committee, had described this as "the Panama equivalent" in his 1992 force-sizing exercise.

- **Overseas presence operations.** The third set of requirements for sizing general-purpose forces was to sustain an overseas presence by U.S. military forces, to protect and advance U.S. interests, and to perform other functions that contributed to security. The BUR planned for some 100,000 troops in Europe and "close to" 100,000 troops in the Pacific/East Asian theater. It also determined that presence needs led to a somewhat higher number of Navy aircraft carriers (11 active and 1 reserve training carrier) than was suggested by warfighting requirements alone, for which 10 carriers would have been adequate.

- **Deterrence of WMD attack.** Deterring WMD attacks against U.S. territory and forces was not a major driver of conventional force structure. However, several of the specialized "new initiatives" proposed by the BUR—especially cooperative threat reduction, counterproliferation, and defense/military partnerships with the former Soviet Union—and the increased emphasis on theater missile defense sought to address this mission area.

Force Options. The BUR's focus was squarely on sizing general-purpose forces and on determining the nature of "force enhancements" necessary for fighting major regional contingencies.[13] To accomplish this goal, the BUR combined the building blocks just discussed with alternative "strategies" to develop force structure options for general-purpose forces.

The range of strategies assessed in the BUR specified successively more demanding environments for U.S. forces: (1) win one MRC; (2) win one MRC with a hold in the second ("win-hold-win"); (3) win two nearly simultaneous MRCs; and (4) win two nearly simultaneous MRCs plus conduct smaller operations. The force structures associated with each strategy are described in Table 3.2.

Of the four options reported above, only the second and third appear to have been given serious consideration by policymakers.[14] In late

---

[13] The BUR's and QDR's focus on two nearly simultaneous MRCs—and their failure to fully reckon the impact of peacetime engagement activities on warfighting readiness and strategic risk—effectively led to what many have termed a "two-conflict strategy." While an oversimplification, this term harbors an important truth.

[14] See U.S. Congress, Senate, "Force Structure Levels in the Bottom-Up Review," *Department of Defense Authorization for Appropriations for Fiscal Year 1995 and the*

Table 3.2

Alternative Force Options Considered in the BUR

Strategy	1: Win One MRC	2: Win One MRC with Hold in Second	3 (BUR Force): Win Two Nearly Simultaneous MRCs	4: Win Two Nearly Simultaneous MRCs Plus Conduct Smaller Operations
Army				
AC divisions	8	10	10	12
RC	6 equivalent divisions	6 equivalent divisions	15 enhanced-readiness brigades	8 equivalent divisions
Navy				
CVBGs[a]	8	10	11 1 RC/training carrier	12
Marines				
AC brigades	5	5	5	5
RC division	1	1	1	1
Air Force				
AC TFWs	10	13	13	14
RC TFWs	6	7	7 Plus force enhancements	10

[a]CVBG = Carrier Battle Group.

May and early June 1993, various DoD sources signaled that the second strategy, win-hold-win, was emerging as the preferred one.[15] However, the resulting criticism from Congress and in the press,[16]

Future Years Defense Program, Washington, D.C.: Government Printing Office, March 9, 1994, pp. 687–753.

[15]There were reportedly three strategies and force structures under consideration in late May that track with options one, two, and four. See Michael Gordon, "Cuts Force Review of War Strategies," *New York Times*, May 30, 1993, p. 16. By the time of his testimony before the Senate Armed Services Committee on June 17, 1993, Deputy Defense Secretary William Perry described three strategies: one that would be capable of one major regional conflict; one that could deal with two simultaneous MRCs; and an intermediate case involving two nearly simultaneous MRCs. He also suggested that win-hold-win was somewhat misleading and that it really was a strategy for "nearly simultaneous" conflicts.

[16]See, for example, Dov S. Zakheim, "A New Name for Winning: Losing," *New York Times*, June 19, 1993, p. 21, and John T. Correll, "Two at A Time," *Air Force Magazine*, Vol. 76, No. 9, September 1993.

together with the concerns articulated by key U.S. allies,[17] led on June 25 to Defense Secretary Les Aspin's public endorsement of strategy (and force structure) three, a strategy and force structure for fighting two nearly simultaneous MRCs. These capabilities were deemed necessary to enhance the probability that a second conflict did not emerge while the United States was already preoccupied with a first conflict—a feature that the win-hold-win strategy lacked.

Force Enhancements Required to Make the Strategy Work. As shown in Table 3.2, strategy and force structures two and three differed very little in their composition. Accordingly, one of the main premises of the strategy and force structure for fighting two nearly simultaneous MRCs was that a number of key force enhancements would need to be undertaken to make the force capable of executing that strategy.[18] These force enhancements aimed to compensate for force levels lower than those of the Base Force through selective investment in core capabilities that could improve the United States' capacity to halt a short-warning attack. Secretary Aspin identified two low-tech enhancements and two high-tech ones as being critical to accomplishing the aim of stopping invading forces as quickly as possible.[19] The low-tech enhancements were increased airlift and prepositioning, and the high-tech enhancements were advanced antiarmor munitions and electronic battlefield surveillance.[20]

Balancing Warfighting and Presence Needs. The BUR also laid down an elaborate logic to ensure the force's ability to disengage from peacetime operations in the event that one or more MRCs were to arise (see Figure 3.2).

---

[17]South Korea (which presumed that it would be put on hold while the United States prosecuted its first MRC in Southwest Asia) reportedly expressed concerns about the strategy.

[18]As described in testimony by Deputy Secretary Perry in the summer of 1993, the primary point of the BUR was not so much to select a particular strategy as to indicate the connection between strategies and military scenarios and the consequent actions that needed to be taken to modernize the force, provide infrastructure, and size budgets.

[19]See Secretary of Defense Les Aspin's remarks at the National Defense University graduation, Fort McNair, Washington, D.C., June 16, 1993, as prepared for delivery.

[20]As noted in the last chapter, some of these "enhancements" had in fact already been planned and/or programmed.

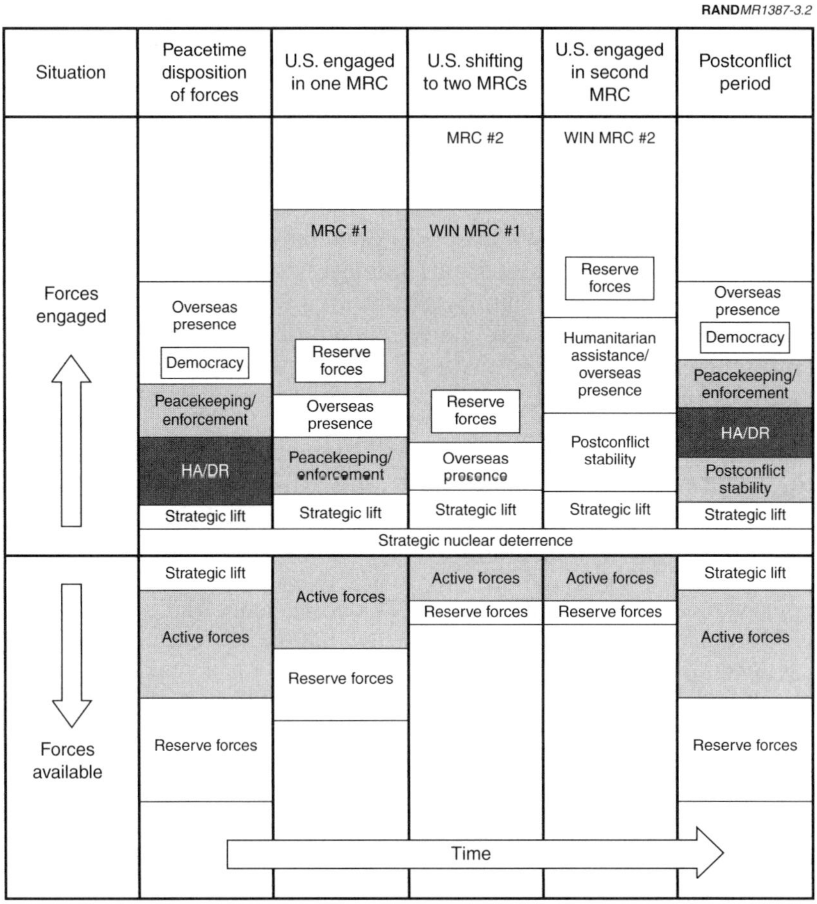

The Bottom-Up Review: Redefining Post–Cold War Strategy and Forces 51

Situation	Peacetime disposition of forces	U.S. engaged in one MRC	U.S. shifting to two MRCs	U.S. engaged in second MRC	Postconflict period
				MRC #2	
				WIN MRC #2	
		MRC #1	WIN MRC #1		
Forces engaged ↑	Overseas presence			Reserve forces	Overseas presence
	Democracy	Reserve forces		Humanitarian assistance/ overseas presence	Democracy
	Peacekeeping/ enforcement	Overseas presence	Reserve forces		Peacekeeping/ enforcement
	HA/DR	Peacekeeping/ enforcement	Overseas presence	Postconflict stability	HA/DR
	Strategic lift	Strategic lift	Strategic lift	Strategic lift	Postconflict stability
					Strategic lift
			Strategic nuclear deterrence		
Forces available ↓	Strategic lift	Active forces	Active forces	Active forces	Strategic lift
	Active forces		Reserve forces	Reserve forces	Active forces
		Reserve forces			
	Reserve forces				Reserve forces

Time →

Figure 3.2—The BUR's View of Conflict Dynamics

As Figure 3.2 shows, this logic dictated that during peacetime, U.S. forces were expected to be conducting overseas presence operations (including support for democracy), peacekeeping and peace enforcement operations, and humanitarian assistance and disaster relief (HA/DR) operations. Such operations were expected to engage a substantial fraction of the force structure.

In the event that the United States became engaged in an MRC, the BUR held that humanitarian, disaster relief, and democracy operations would be dropped while overseas presence operations would shrink; peacekeeping and peace enforcement operations were, however, expected to continue at their peacetime levels. With a second MRC, the United States would focus on winning the first MRC, while participation in peacekeeping and peace enforcement operations would be suspended; other overseas presence operations would continue at the reduced levels of the one-MRC case. With the successful conclusion of the first MRC, postconflict stability operations would begin in that theater for an indeterminate period, and with the win in the second MRC, similar operations would begin in that theater. These operations were then expected to continue into the postconflict period until they were concluded, whereupon the United States would return to its peacetime disposition of forces.

Despite this attention to presence and to peace and relief operations, during the DoD press conference on the BUR, Chairman Powell was emphatic in explaining that the BUR force was justified primarily in terms of its warfighting ability:

> Let me begin by giving a little bit of a tutorial about what an armed forces is all about. Notwithstanding all of the changes that have taken place in the world, notwithstanding the new emphasis on peacekeeping, peace enforcement, peace engagement, preventive diplomacy, we have a value system and a culture system within the armed forces of the United States. We have this mission: to fight and win the nation's wars. That's what we do. Why do we do it? For this purpose: to provide for the common defense. And who do we do it for? We do it for the American people. We never want to lose sight of this ethic, we never want to lose sight of this basic underlying principle of the armed forces of the United States. We're warriors. And because we are warriors, because we have demonstrated time and time again that we can do this for that purpose for the American people, that's why you have armed forces within the United States structure.
>
> Now at the same time, because we are able to fight and win the nation's wars, because we are warriors, we are also uniquely able to do some of these other new missions that are coming along—peacekeeping, humanitarian relief, disaster relief—you name it, we can do it. And we can modify our doctrine, we can modify our strategy, we can modify our structure, our equipment, our training, our leadership techniques, everything else to do these other missions,

but we never want to do it in such a way that we lose sight of the focus of why you have armed forces—to fight and win the nation's wars.[21]

In carrying out the budget cuts, administration policymakers hoped to cut defense spending without provoking questions about their commitment to the nation's defense. Paradoxically, this resulted in more modest cuts in force structure in the BUR than had been advocated by Secretary Aspin when he had been House Armed Services Committee chairman, as well as deeper cuts in defense resources than had been advocated in the campaign. Another result was that a strategy—first win-hold-win and then a strategy for fighting two nearly simultaneous major wars—was overlaid on a force structure that was justified by the CJCS in terms of warfighting ability but would instead become preoccupied with operations in support of the administration's still-crystallizing strategy of "engagement and enlargement." A final result was a gap between publicly stated expectations of $104 billion in defense savings and the $17 billion that was privately expected by OSD policymakers.

Assignment of Forces for Overseas Presence. While the capability to prosecute two nearly simultaneous MRCs was the principal yardstick for sizing U.S. forces, it was the BUR's assessment of peacetime overseas presence that defined the logic of its assignment of active forces to various regions (see Table 3.3).[22]

As shown in Table 3.3, overseas presence needs dictated that roughly 25 percent of active Army divisions, slightly more than 40 percent of USAF TFWs, one-third of the active MEFs, and nearly 25 percent of Navy aircraft carriers be deployed outside the continental United States. Nevertheless, the Air Force did not actively press the case that, as with the Navy carriers, presence needs and support to contingencies should also be considered in determining the number of TFWs in the force structure.

---

[21] Department of Defense, news conference on the DoD Bottom-Up Review, September 1, 1993.

[22] Although the assessment provided the logic, it appears that the BUR's assessment of overseas presence was intended not so much as an assignment of forces to regions as a public vindication before the United States and its allies of the levels of forces that were being retained.

Table 3.3

Assignment of Forces in the BUR, October 1993

Force Package	Army Divisions	USAF TFWs	USMC MEFs	Navy Carriers[a]
Forward				
Atlantic				
Europe	2.0	2.3		0.7
Pacific				
Japan	0.11[b]	1.4	1.0	1.0
South Korea	0.66[c]	1.0	(0.66)[d]	
Persian Gulf	—[e]	—[e]	(0.33)[f]	0.7–1.0[g]
Total	2.66	4.7	1.0	2.4–2.7
Contingency				
U.S.				
Active	7.33	8.3	2.0	7.3–7.6
Reserve	5.0+	7.0	1.0	1.0[h]
Total	12.33	15.3	3.0	8.3–8.6
Total	15.0+	20.0	4.0	12.0
Active	10.0	13.0	3.0	11.0
Reserve	5.0+	7.0	1.0	1.0[h]

SOURCES: Les Aspin, *Report on the Bottom-Up Review*, Washington, D.C., October 1993, pp. 23–24, and authors' estimates.

[a] Eleven active plus one reserve carrier capable of sustaining full-time presence in one region and presence in two other regions 70 percent of the time.

[b] Army Special Forces battalion in Okinawa, not scored against division count.

[c] The BUR plan called for ultimate reduction to one brigade.

[d] Two brigade-size MEFs (two maritime prepositioning squadrons [MPSs]) available for the MRC in Northeast Asia.

[e] Land-based Army and Air Force forces to be rotational only.

[f] One MPS for a brigade-size MEF available for the MRC in Southwest Asia.

[g] "Tether" carrier, supplemented by Middle East Force (MIDEASTFOR) ships and a rotational Amphibious Readiness Group (ARG).

[h] Reserve training carrier.

Force Structure and Manpower. The proposed manpower and force structure changes to achieve the BUR force over the FY 1995–1999 program included reductions to manpower of some 160,000 active personnel and 115,000 civilians.

For the Air Force, the BUR proposed to reduce the number of TFWs by an additional 6.5 wings below planned Base Force levels, including 2.25 active and 4.25 reserve wings, and to set the number of bombers at 184, 114 of which would be scored as part of the strategic nuclear force.[23] Army divisions were to be reduced by two below the planned Base Force number while maintaining five or more reserve divisions.[24] Naval forces were to be reduced by 55 surface ships and submarines, including the nominal cutting of one aircraft carrier, thereby reducing the carrier force level from 12 to 11 plus one reserve training carrier. One active and one reserve Navy air wing would also be cut. The Marines were to see an increase in planned end strength from the Base Force plan of 159,000 to 174,000.

In short, BUR policymakers stated their aim to accomplish with a smaller force what the Base Force could do only with great difficulty, and placing it near its breaking point—providing a capability to fight two nearly simultaneous major conflicts. Furthermore, this force would also be employed in peace, humanitarian, and other non-warfighting operations to a much greater degree than had been envisioned in the Base Force and was said to require $104 billion less than the Bush baseline had provided for the Base Force. This tenuous balance between strategy, forces, and resources struck in the BUR would set the stage for many of the problems encountered over the years that followed.

The View from the Air Force. Another key assumption of BUR policymakers had an important effect on force structure, particularly that of the Air Force. As was noted in Chapter Two, the Base Force had been predicated in part on the assumption implied in the Army's review of its warfighting doctrine, "Airland Battle," that future wars would involve significant clashes of armor against armor.

The Gulf War's reliance on air power to destroy Iraqi strategic targets, defeat Iraqi fielded forces, and create the conditions for the successful ground offensive raised the question of whether aerospace power would also be used to establish air supremacy and defeat mechanized ground forces in future conflicts, and whether the future composition of U.S. forces should instead favor aerospace power over

---

[23]Strategic nuclear forces were not examined in detail in the BUR but were addressed in the Nuclear Posture Review. The B-2s were, however, to be capable of both strategic nuclear and conventional missions.

[24]The details of Army reserve-component forces would be worked out in the fall of 1993.

heavy ground forces. The BUR (and later the QDR) seems to have leaned toward such an alternative view—and accordingly favored a number of force enhancements to improve the ground attack capabilities of bombers and tactical aircraft to enhance U.S. capabilities to halt an enemy offensive.

Resources

When the BUR spending plan for the FY 1995–1999 FYDP is compared with the Bush baseline for the same years (Table 3.4), the differences between the two plans emerge clearly.

The Bush administration's final spending plan had anticipated a reduction of nearly 26 percent in DoD budget authority from FY 1990

Table 3.4
The BUR's Long-Range Forecast for DoD

			Estimate					
	1990	1993[a]	1995	1996	1997	1998	1999	1995–1999
Bush baseline								
BA ($B)[b]	293	259	257	261	264	270	273	1325
BA (FY 1995 $B)	339	280	263	261	257	256	252	1187
Percent real change:								
From FY 1990			–22.6	–23.0	–24.1	–24.5	–25.8	
From FY 1993			–3.0	–3.6	–4.9	–5.4	–7.0	
BUR plan								
BA ($B)	293	259	249	242	236	244	250	1221
BA (FY 1995 $B)	339	280	249	236	224	225	224	1157
Percent real change:								
From FY 1990			–26.6	–30.5	–34.0	–33.7	–33.9	
From FY 1993			–8.0	–12.9	–17.3	–16.9	–17.2	
Reduction								
BA ($B)			8	19	28	26	23	104
BA (FY 1995 $B)			14	25	34	31	28	131

SOURCES: Les Aspin, *Report on the Bottom-Up Review*, Washington, D.C., October 1993, p. 108, and DoD Comptroller, *National Defense Budget Estimates, FY 1995*, Washington, D.C., March 1994.

[a]The estimate for FY 1993 is from DoD Comptroller, *National Defense Budget Estimates FY 1994*, Washington, D.C., May 1993.

[b]BA = Budget Authority.

to 1999, with 7 percent of those reductions taken after FY 1993. By comparison, the budget proposed in the BUR anticipated a 34 percent reduction by the end of the FY 1990–1999 period, with reductions of an additional 17 percent after FY 1993. Thus, the BUR anticipated approximately 8 to 10 percent in additional reductions beyond the Bush administration's baseline spending plan.

The BUR reported that the administration had set a target of $104 billion in savings for the FY 1995–1999 budget and program and detailed a total of $91 billion in estimated savings, leaving a shortfall of $13 billion.[25] Privately, however, OSD policymakers reportedly anticipated only $17 billion in savings. In any event, the cuts that were to be made to achieve the $91 billion savings target were to come predominantly from modernization accounts ($53 billion) but were also to be derived from force structure and infrastructure ($43 billion). This $96 billion in reductions was to be offset by a $5 billion increase in funding for new initiatives, including peacekeeping and peace enforcement operations.

Defense Priorities. In the face of further anticipated budget cuts, the BUR undertook only selected modernization to improve capabilities against enemy mechanized forces through programs such as precision-guided munitions and improved surveillance. The BUR sought to address the problem of the aging aircraft inventory and the bow wave associated with the procurement of the next generation of aircraft (see Figure 3.3) by canceling the A/F-X and Multirole Fighter (MRF),[26] terminating production of the F-16 after FY 1994 and the F/A-18C/D after FY 1997, and proceeding with the F-22 and the F/A-18E/F, albeit at reduced quantities.[27] The F-22 was also to be given a precision ground attack capability at the outset of its production,

---

[25]With OMB's updated Mid-Session Review revision of inflation estimates, the FYDP shortfall grew to $20.5 billion, and military and civilian pay raises generated an additional shortfall of roughly $11.4 billion, leading to a total shortfall of $31.5 billion. OMB increased the DoD budget over the FYDP to cover the pay raises but not the multiyear inflation bill, leaving an estimated shortfall of some $20 billion.

[26]In its place, a Joint Advanced Strike Technology (JAST) program was established that would seek to develop common components and subsystems that could be used in building a number of service-specific fighter/attack aircraft platforms.

[27]The size of the F-22 buy fell by roughly two wings (210 aircraft, including backup aircraft inventory [BAI], attrition reserve, and the like), from 648 to 438 aircraft.

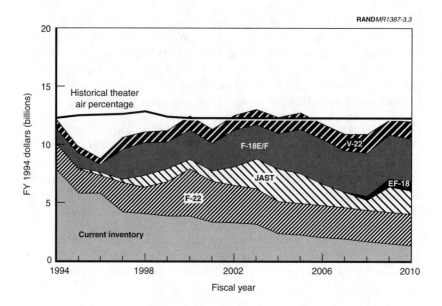

Figure 3.3—The BUR's Plan for Eliminating the Bow Wave in the Theater Air Program[28]

thus providing a multirole capability to increase the aircraft's utility and cost-effectiveness.

The BUR recommended a total of $5 billion over the FY 1995–1999 period to support four new policy initiatives: cooperative threat reduction, counterproliferation, expanded contacts with the former Soviet Union to create a defense/military partnership, and global cooperative initiatives. The final initiative included peacekeeping and peace enforcement, humanitarian assistance and disaster/famine relief, and the promotion of democracy through military-to-military contacts.[29]

---

[28] The labeling in this figure is as it appeared in Aspin, *Report on the Bottom-Up Review*.

[29] BUR policymakers anticipated that future peace operations would be paid for out of a special account. It was not until the Overseas Contingency Operations Transfer Fund (OCOTF) late in the decade, however, that such a mechanism would be created. The result was a reliance on emergency supplementals and the annual appropriation process.

IMPLEMENTING THE BUR

Strategy

The unsettled international environment and the administration's promotion of peace operations as an appropriate response to this instability and conflict resulted in commitments throughout the 1993–1998 period that were, from a historical perspective, more frequent, larger, and of longer duration than had been seen in the past (see Figure 3.4). The result was by some accounts a commitment to smaller-scale contingencies (SSCs) that began to approximate the requirements of a single MTW,[30] together with growing congressional and other concern about the potential impact of such SSCs on readiness for warfighting.

Somewhat less obvious is that these operations were also unlike most past operations in the sense that the United States was making long-

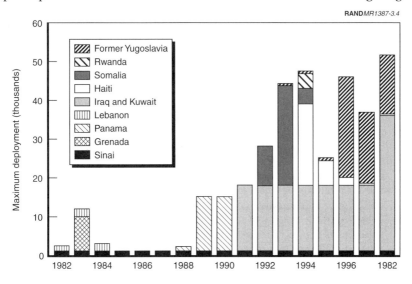

Figure 3.4—Maximum Deployment in Peace Operations, 1982–1998 (CBO)

---

[30] See Michael C. Ryan, *Military Readiness, Operations Tempo (OPTEMPO) and Personnel Tempo (PERSTEMPO): Are U.S. Forces Doing Too Much?* Washington, D.C.: Congressional Research Service Report 98-41F, January 14, 1998.

term commitments to operations in relatively austere out-of-area locales that would require sustainment through rotation—this at a time when the United States continued to reduce its overseas force levels and shift its posture to emphasize U.S.-based contingency forces. In combination with force structure and manpower reductions, the result of these commitments was to greatly increase operational and personnel tempos over the period.

As indicated by congressional reaction to the BUR and hearings on readiness over the 1994–1998 period, this increased operational tempo across the force underscored congressional concerns that the forces were insufficient to underwrite the emerging strategy of engagement and enlargement, that the available resources were insufficient to maintain the health of the force, and that readiness for warfighting would ultimately suffer.[31]

The View from the Air Force. Air Force leaders expressed some disappointment that the BUR had failed to tackle the issue of roles and missions and the restructuring of U.S. forces. Air Force Chief of Staff Merrill A. McPeak argued that the issue, while important, seemed constantly to be trumped by other, more critical issues on the defense secretary's plate:

> Every morning when the SECDEF comes to work, he faces a problem all leaders face: How to distinguish between what's important and what's critical. What's important is that we organize the Nation's defenses properly. That's roles and missions. What's critical is Bosnia. He can't ignore the critical problems in order to pay attention to the important ones. So, I don't think you can rely on the Secretary of Defense or the Chairman of the Joint Chiefs. They are simply tied up with other problems that are seen as more pressing because they are in the headlines every day. Roles and missions is probably more important, but it's not on the 6:00 news every day. It can be ignored, and accordingly it will be, unless the President wants to take an interest in it.[32]

---

[31] As will be described in the next chapter, until the early fall of 1998, most civilian and military leaders publicly rejected the argument that readiness had deteriorated.

[32] George M. Watson and Robert White, end-of-tour interview with General Merrill A. McPeak, Air Force Chief of Staff, conducted at the Pentagon, November 28 and December 15 and 19, 1994.

Nevertheless, there were indications that the president had taken an interest in the subject:

> By the way, the President made a wonderful speech about this subject [of roles and missions] in August of 1991 when he was campaigning in Los Angeles. He talked about what was needed doing, and he had a lot of it exactly right. I urge you to read that speech sometime. It was very good. The thrust was that we must downsize the Armed Forces, and it is a disservice to the Nation if we simply do it as "Cold War–minus." He accused the Bush administration of taking this approach. The so-called "Base Force" was kind of a Cold War–minus approach: just cut everybody 30 percent, walk away from it, and wash your hands. What the President said was, "All that does is make us 30 percent weaker than we were before, and that's not good enough. We have to rebuild the Armed Forces, eliminate duplication and overlap, and so on. Then we can cut it and maybe be as strong or stronger than we were before because we have rethought the problem of who is going to do what in a more imaginative way." That speech was exactly correct. What happened was that the new Administration came in and they didn't do that. They did the so-called "Bottom-Up Review," which was Cold War minus-minus. They took the Base Force down another 30 percent and didn't redistribute any of the jobs in any way whatsoever, let alone more efficiently.
>
> In my view, the President must have been disappointed with the Bottom-Up Review, although I was at the White House when Secretary Aspin briefed it to the President. All the Chiefs were there. The President said "This is brilliant work. It is exactly what we needed." I kept watching him to see if he was serious or if he was just being a nice guy. It was hard for me to tell whether he really believed the Bottom-Up Review was a brilliant piece of work, but those were exactly the words he used. In any case, he certainly did not follow the prescription that he had laid down in his campaign, which was exactly right. What we should have done is what I would call a Wall-to-Wall Review, as opposed to a Bottom-Up Review. A Wall-to-Wall Review would look at the range of tasks we are doing here and decide how to do each of them best.[33]

---

[33] Ibid.

Notwithstanding this disappointment, as it had with the Base Force, the Air Force embraced the new strategy and its emphasis on long-range aerospace power, including long-range conventional bombers, strategic mobility, enhanced surveillance and targeting, and precision-guided attack. The new strategy also placed unprecedented demands on the Air Force in servicing contingency operations over the period. The number of deployed aircraft remained at a level substantially higher than before the Gulf War as a result of the need to service the operations in northern and southern Iraq (see Figure 3.5). Modest increases in the number of aircraft in contingency operations can be seen thereafter as additional commitments accumulated, particularly in Bosnia. In general, more than 200 USAF aircraft were deployed fairly consistently to contingency operations throughout the 1993–1998 period, although occasional peaks of 350 or more aircraft were also seen.

With force structure having twice been traded for modernization, the die was effectively cast, and the Air Force faced the continued prospect of underwriting a more ambitious strategy with a smaller force. While some of the resulting stresses would be mitigated somewhat by several Air Force post-BUR innovations,[34] by March 1997 Air Force Chief of Staff Ronald Fogleman reported that Air Force operational tempo was four times that demanded prior to the fall of the Berlin Wall.

Force Structure and Manpower

The force targets and the execution of force structure reductions for the BUR generally appear to have gone as planned, with most force structure goals achieved by FY 1996–1998[35] and with some goals

---

[34] See the posture statement of General Fogleman, presented in testimony before the House National Security Committee, March 5, 1997. Among the Air Force innovations in the post-BUR period were the Global Sourcing Conference in 1995, which better balanced demands across major commands, and the air expeditionary force concept in 1996, which reorganized the Air Force to better service ongoing commitments while retaining a capability to respond to crises.

[35] Nominal force structure goals were met in the following fiscal years: active Army divisions (FY 1996); Navy ships (FY 1997–1998); Air Force reserve-component TFWs (FY 1996). By 1997, the enhancement of Army separate brigades was also almost completed. See Department of Defense, *Report of the Quadrennial Defense Review*, Washington, D.C., May 1997, p. 32.

The Bottom-Up Review: Redefining Post–Cold War Strategy and Forces 63

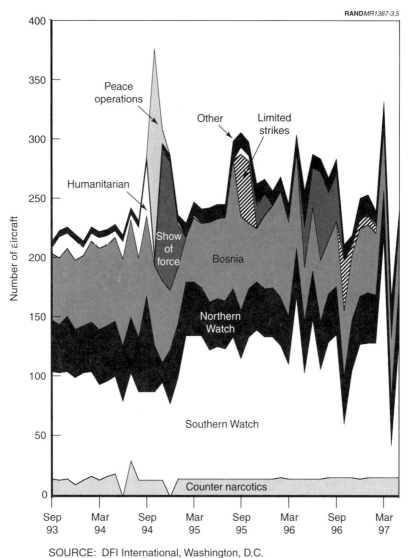

Figure 3.5—USAF Aircraft in Contingency Operations, 1993–1998

actually achieved as early as FY 1994, even before BUR implementation had formally begun.[36] In the final months of 1993, however, several important changes to the force levels appear to have been specified in the BUR. Specifically, the number of Army reserve divisions and brigades was amended in December 1993, leading to a force goal of eight Army National Guard (ARNG) divisions and 18 ARNG brigades, 15 of which would be enhanced-readiness brigades.

The View from the Air Force. For the Air Force, the number of planned Primary Aircraft Authorization (PAA) bombers was reduced for the FY 1995 defense budget and plan from "up to 184" to a new goal of 140—a target that was reached by 1998, although the composition differed somewhat from what had been planned. A decision was also made to reduce the total number of Navy ships from 346 to 331 by FY 1999.

For the Air Force, force structure reductions fell unevenly across the force both during the transition year of FY 1994 and during the implementation of the BUR decisions. Figure 3.6 shows the cumulative reductions to various elements of force structure and infrastructure associated with the Base Force (FY 1990–1993), the transition year of FY 1993–1994, and the BUR (FY 1994–1997).

The FY 1994 transition year can be seen to have contributed substantial reductions beyond the Base Force, including reductions to Air Force active and reserve TFWs, reserve total aircraft inventory (TAI), active bombers, and infrastructure. During execution of the BUR from FY 1995 to FY 1997, the greatest relative reductions focused on reserve TFWs, reserve TAI, active strategic airlift, and infrastructure. And as noted above, with the FY 1995 defense budget and the FY 1995–1999 defense program, the number of long-range bombers was reduced from up to 184 to 140 by FY 1999.[37]

---

[36]These included reserve-component Army divisions, Air Force TFWs, Navy aircraft carriers, and Marine end strength.

[37]Of these, 48 were to be B-52H bombers equipped to carry both nuclear-armed air-launched cruise missiles (ALCMs) and conventional weapons, 72 were to be B-1B (all to be converted to conventional weapons only by 1998), and 20 were to be B-2s with conventional and nuclear weapon delivery capabilities. See Department of Defense, *Annual Report to the President and Congress*, Washington, D.C., January 1994, p. 27.

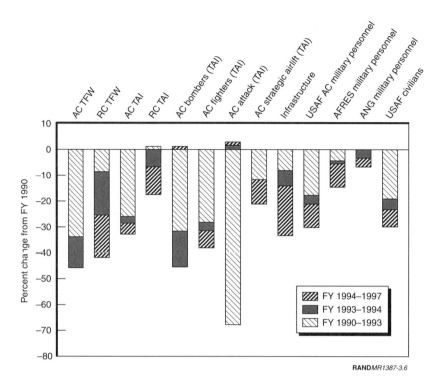

Figure 3.6—USAF Force Structure and Manpower Reductions, FY 1990–1997

Infrastructure

A 1995 round of the BRAC Commission further reduced domestic bases by some 20 percent DoD-wide. During this time frame, Air Force infrastructure fell by a relatively low 14 percent.[38] By FY 1998, the total number of Air Force installations worldwide—including major and minor installations and support sites—had declined by 233 installations, or some 45 percent, with most of the decline oc-

---

[38]From FY 1993 to 1998, USAF infrastructure spending as a fraction of total spending fell from 44 to 42 percent, where it had been in FY 1990. See U.S. Air Force, *Air Force Strategic Plan, Vol. 2*, Table 2.B.3, "Total Infrastructure Spending."

curring during the BUR years of FY 1995–1998.[39] As a result of congressional dissatisfaction with the selection process for the last round of base closures, however, no additional BRAC rounds were authorized; instead, Congress essentially decided to revisit the question after a new administration was in place in 2001.

Modernization

Modernization decisions reported in the BUR resulted in a number of program terminations in the FY 1995 budget and program, including the A/F-X, EA-6B remanufacture, F-16, CH-53, SH-60B/F/H, MRF, Follow-on Early Warning System (FEWS), Spacelifter, and LANDSAT satellite.[40] Other planned modernization programs did not achieve their targets, including the following:

- **Theater air program.** Apparently as a result of cuts to modernization accounts occasioned by higher-than-expected O&S costs,[41] the theater air program was not executed as planned, resulting in the creation of another procurement bow wave that was left to the 1997 QDR to resolve.[42] Similarly, the initial oper-

---

[39] According to the USAF's *Statistical Digest*, approximately 278 Air Force installations were closed between FY 1995 and FY 1998. See Assistant Secretary of the Air Force (Financial Management and Comptroller), *United States Air Force Statistical Digest*.

[40] Unclassified extract from Joint Chiefs of Staff, *Joint Military Net Assessment*, Washington, D.C., 1994, p. 3-8.

[41] For example, our comparison of the December 1993 and December 1998 Selected Acquisition Reports (SARs) for the F-22 showed that actual spending on the F-22 over FY 1996–1998 was $2.1 billion less than planned, and rather than beginning in FY 1997, acquisition of the F-22 did not begin until FY 1999. The DoD would later observe in the 1997 QDR that each new defense program since the BUR had had to postpone the previous year's plan to increase procurement spending, and these postponements reflected the importance that the DoD attached to current spending on readiness. Funding originally planned for procurement was spent instead to meet day-to-day operating expenses, a phenomenon the DoD referred to as "migration" of funding. See U.S. General Accounting Office, *Future Years Defense Program: DoD's 1998 Plan Has Substantial Risk in Execution*, Washington, D.C., GAO/NSIAD-98-26, October 1997, p. 21. Between FY 1996 and 1999, the result was an estimated $45.8 billion in unrealized procurement spending. See U.S. General Accounting Office, *DoD Budget: Substantial Risks in Weapons Modernization Plans*, Washington, D.C., GAO/T-NSIAD-99-20, October 8, 1998, p. 4. Our analysis of the Aircraft Procurement, Air Force accounts for the BUR years suggests that in three of the years in the FY 1995–1999 period, actual spending was more than $225 million below planned spending.

[42] This will be discussed in more detail in the next section.

ational capability (IOC) for the F-22 has slipped from FY 2003, as anticipated in the BUR, to FY 2005. In addition, the planned "EF-X" was never fielded to replace the aging F-4Gs, EA-6Bs, and EF-111s, which led to a decision to configure F-16s for the lethal Suppression of Enemy Air Defenses (SEAD) mission (F-16CJs) and to continue using the fleet of Navy and Marine Corps (and joint) EA-6Bs.

- **Enhancing long-range bombers' conventional capabilities.** Despite some program slippage, the continued modification of long-range bombers to improve their conventional capabilities generally appears to have been executed as planned, with B-1Bs, B-2s, and B-52Hs receiving planned upgrades to incorporate the Joint Direct Attack Munition (JDAM), Wind-Corrected Munition Dispenser (WCMD), Joint Air-to-Surface Standoff Missile (JASSM), and other precision munitions. Block E upgrades to the B-1B began in FY 1996,[43] and Block F upgrades began in FY 1997. A Block 30 upgrade for the B-2 was begun in FY 1997, and an advanced weapon integration program for the B-52H was begun in FY 1996.[44]

- **Enhanced precision-guided munitions.** As part of the package of force enhancements, the BUR advocated continued support to build on existing stocks of precision-guided munitions and increased support for the acquisition of new all-weather precision-guided munitions. Although most of these systems were pursued after the BUR, no procurement of these systems appears to have taken place during FY 1993–1995.

- **Battlefield surveillance.** The BUR supported a buy of 20 JSTARS, although only 19 aircraft were ultimately approved.[45]

---

[43] Block E included a capability for the WCMD, JASSM, and Joint Standoff Weapon (JSOW); Block D upgrades, which provided a capability to deliver JDAMs, had begun in FY 1994.

[44] This program provided B-52Hs with a capability to deliver JDAM, WCMD, JSOW, and JASSM.

[45] See U.S. Congress, Senate, "Force Structure Levels in the Bottom-Up Review," p. 738. On September 25, 1996, the Under Secretary of Defense for Acquisition approved full-rate production of JSTARS with a total planned quantity of 19 production aircraft.

- **Enhanced mobility.** As had the Base Force policymakers, the BUR embraced the 1992 MRS's findings that 120 C-17s' worth of airlift capacity and additional large, medium-speed roll-on/roll-off (LMSR) ships were needed,[46] and that prepositioning—including additional afloat prepositioning ships (APSs)—needed to be enhanced. Of these mobility enhancements, only the C-17 and prepositioning goals seem to have been met.[47]

In the end, the planned ramp-up in modernization continued to be deferred and was perhaps the central issue to be readdressed by the 1997 QDR and its proposed strategy of "shape, respond, and prepare now."

Resources

Spending Plans. As shown in the updated "pitchfork" chart in Figure 3.7, the FY 1995 defense spending plan accelerated the budget re-

---

[46] Although the capacity requirement was reaffirmed, the C-17 buy quantity was reduced to 40 aircraft at the time of the BUR pending the resolution of a number of serious problems that plagued the program at the time. With the resolution of these problems, the buy quantity was restored to 120 aircraft in late 1995.

[47] In 1994, the Mobility Requirements Study, Bottom-Up Review Update (MRS BURU) established an airlift goal of 49.7 million ton-miles per day (MTM/D) by FY 2001. Although the C-17 buy has gone according to plan, overall capacity has been affected by C-5 maintenance problems and by an accelerated C-141 drawdown. In March 1999, General Charleston Robertson, the U.S. Transportation Command's Commander in Chief, testified before the House Armed Services Committee that the airlift fleet was 5.43 MTM/D short of that goal. See statement of General Charleston T. Robertson, Jr., USAF, Commander in Chief, U.S. Transportation Command, before the House Armed Services Committee, March 22, 1999. By 2001, the GAO was reporting that the military wartime airlift capability shortfall was 5.76 MTM/D, or nearly 20 percent of the overall requirement, while the tanker refueling shortfall was 30 percent (total refueling capacity) to 39 percent (total refueling aircraft). See U.S. General Accounting Office, *Military Readiness: Updated Readiness Status of U.S. Air Transport Capability*, Washington, D.C., GAO-01-495R, March 16, 2001, p. 11.

The MRS BURU also established a goal of 19 LMSR ships for prepositioning and surge sealift by 2001. By 1997, deliveries were behind schedule, and by FY 1999 a total of only 12 ships had been acquired. See U.S. General Accounting Office, *Strategic Mobility: Late Deliveries of Large, Medium-Speed Roll-on/Roll-off Ships*, Washington, D.C., GAO/NSIAD-97-150, June 1997. In his March 1999 testimony, General Robertson reported that USTRANSCOM forecast that a surge sealift shortfall of 400,000 square feet would remain by FY 2001.

The Bottom-Up Review: Redefining Post–Cold War Strategy and Forces 69

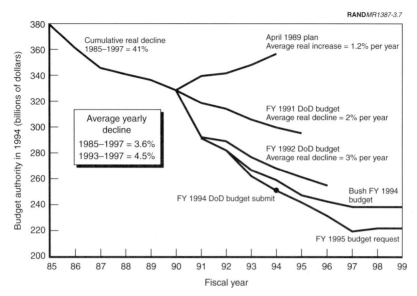

SOURCE: Joint Chiefs of Staff, *Joint Military Net Assessment*, Washington, D.C., August 13, 1994.

Figure 3.7—"Pitchfork" Chart, Circa 1994[48]

ductions begun during the Base Force period and in the FY 1994 defense budget and program.[49]

The long-range defense spending plan associated with the FY 1994 budget and program submitted in spring 1993 appears to have created fairly binding—and tight—top-level constraints on the strategy and force planning done under the BUR; the top lines were virtually identical.[50] The tightness of the program resulted in a number of

---

[48]The reader will note that there are two FY 1994 budgets—one prepared by the outgoing Bush administration and one by the new Clinton administration—and that the FY 1994 and FY 1995 Clinton administration spending plans were nearly identical. Additionally, it is worth mentioning that budget numbers from the Joint Staff are not reviewed by the OSD Comptroller and could contain small inaccuracies.

[49]See Figure 2.4 for an earlier version of the "pitchfork" chart, circa 1992.

[50]As described earlier, the strategy and force planning were said not to have been fiscally constrained. Nevertheless, BUR planners were able to identify only $91 billion in

risks in its execution. These constraints, along with the difficulty of achieving unrealistic savings targets,[51] were exacerbated by the unwillingness of Congress to approve the administration's pay freeze in 1993; instead, Congress mandated a 2.2 percent military pay raise as well as a locality pay raise for government service employees, which led to an estimated $11.4 billion in additional costs.[52] Although OMB agreed to provide additional funding to cover the pay raise, it was unwilling to add funds to cover higher-than-expected inflation estimates. The result was a defense program for FY 1996–1999 that by all appearances was over budget by roughly $20 billion.

Priorities. Although the FY 1994 budget and program had provided programmatic detail for FY 1994 only, it made clear that readiness-related O&M spending would be kept at high levels; that spending on research and development would expand; and that most of the cuts to achieve administration savings goals would come from force structure and manpower reductions and from reductions to procurement accounts. This priority would also apply to subsequent budgets.

As a percentage of total DoD budget authority, the allocation of resources generally supported these priorities: O&M spending was slated to increase from 27 to 32 percent of total DoD budget authority, RDT&E to increase from 10 to 12 percent, and procurement to fall from 27 to 21 percent (see Table 3.5).

In absolute terms, however, the picture was a bit murkier. Savings from force structure and manpower reductions generally were to be realized through a reduction in spending on military personnel over FY 1994–1999 of nearly $41 billion from the Bush levels. Further, the high priority given to readiness resulted in only slight declines to planned O&M funding over FY 1994–1999. However, in relation to the Bush program, both procurement and RDT&E would decline even more than spending on military personnel over the FY 1994–

---

the $104 billion in cuts sought by the administration, leaving $13 billion to be worked out later.

[51] As described earlier, OSD policymakers privately expected about $17 billion in savings, not $104 billion as reported in the BUR.

[52] We assume but have no direct evidence that the services were required to undertake reprogramming to cover their portions of the bill.

Table 3.5

DoD Budget Authority by Title

Account	Historical Average (%)	FY 1995 (%)
Investment		
R&D	10	12
Procurement	27	21
Military construction	2	1
Operations and support		
Military pay	29	29
O&M	27	32
Family housing	1	2
Other		
Retired pay/accrual	3	2
Other	1	1

SOURCE: Joint Chiefs of Staff, *Joint Military Net Assessment*, Washington, D.C., August 13, 1994, p. 3-6.

NOTE: Budget numbers from the Joint Staff are not reviewed by the OSD Comptroller and could contain small inaccuracies.

1999 period: Procurement was to fall by $47.3 billion and RDT&E by $45.6 billion. In short, relative to the Bush program, the FY 1994 and FY 1995 budgets planned substantial reductions to what had nominally been one of the administration's key priorities: research and development.

These cuts to modernization and investment over FY 1995–1999 were substantially larger than those that had been suggested in the BUR: Rather than seeing a total reduction of $53 billion in cuts as described in the BUR, the FY 1995–1999 program anticipated a total of $79.3 billion in cuts to procurement and RDT&E. When the final accounting was done, these cuts would be even higher.

Two other adjustments occurred in the FY 1994 and FY 1995 spending plans. First, the $20 billion shortfall in anticipated savings from the Bush administration's Defense Management Report initiative was considered in the February 1994 budget submission via as-yet-unallocated reductions over FY 1996–1999.[53] Second, a gap had

---

[53]The anticipated savings was $50 billion out of the $70 billion that had been projected, leaving a $20 billion shortfall.

emerged between anticipated program costs and the budget actually afforded since the budget submission; this was addressed through an unallocated increase of $10 billion that was also spread out over FY 1996–1999.

As a result of higher-than-expected spending on military personnel and O&M activities together with failure to achieve all of the nominally planned savings, actual savings turned out to be much smaller than the $104 billion anticipated in the FY 1994, BUR, and FY 1995 plans (see Table 3.6, which describes the actual differences in spending from the Bush administration to the Clinton administration FY 1994–1999 budgets).

Table 3.6 shows that in the end, the administration realized only some $15 billion in savings over its predecessor's budget and six-year program—closer to the $17 billion in savings OSD policymakers were privately said to have expected than to the more than $100 billion reported in the BUR.

The main reason for the nearly $88 billion shortfall was that both military personnel and O&M spending turned out to be much higher than was assumed by the $104 billion in savings.[54] Further, cuts to procurement turned out to be higher—and cuts to RDT&E lower—

Table 3.6

Difference Between Bush and Clinton Budgets for FY 1994–1999

	1994	1995	1996	1997	1998	1999	1994–1999
Personnel	−0.1	+1.6	+1.1	+0.5	−2.3	−3.2	−2.3
O&M	−1.7	+4.9	+5.3	+4.8	+6.7	+13.2	+33.2
Procurement	−6.4	−11.2	−14.4	−13.1	−14.1	−9.3	−68.6
RDT&E	−7.9	−8.0	−5.2	−2.2	−0.5	+1.1	−22.7
Military construction	−0.1	+0.7	+3.3	+2.2	+1.8	+2.4	+10.2
Family housing	−0.5	−0.6	+0.4	+0.4	+0.0	−0.2	−0.4
Other	+5.7	+6.9	+6.3	+7.5	+3.4	+5.9	+35.8
Total DoD	−10.6	−5.5	−3.0	−0.3	−5.3	+9.4	−15.4

SOURCE: Steven Daggett, Congressional Research Service, updated in October 2000 on the basis of updated deflators for FY 1994–1999.

---

[54] Spending in the "Other" category, which includes some spending on readiness-related revolving funds, was also higher than expected.

than suggested by the FY 1995 plan. As shown in Table 3.6, the total reduction in procurement and modernization relative to the Bush program was roughly $90 billion.

In sum, the FY 1995 budget and program that implemented the BUR greatly underestimated the actual costs of reconciling the BUR's ambitious strategy with its reduced force structure while overestimating the savings over its predecessors' spending plans.

It is worth noting that, contrary to what would appear to be conventional wisdom on the subject, the incremental costs of contingency operations accounted for only $17.7 billion to $18.2 billion—well under one-quarter of the $88 billion shortfall.[55] The more than three-quarters of the shortfall that remained was attributable to other causes, including underestimates of program costs, overestimates of savings, and other technical factors.[56] In the end, the BUR strategy and force structure appear to have required a Base Force–size budget—and the result of this mismatch was a recurring need to find ways to bridge the gap between the budgets that were afforded and the actual costs of the defense program.

The typical pattern over the period was that Congress added money to the President's Budget request,[57] and emergency supplementals were used to cover not only the costs of unanticipated contingency operations but also other desiderata.[58] Finally, subsequent spending

---

[55] Nina Serafino of the Congressional Research Service estimated the incremental costs of U.S. commitments to peace operations over the FY 1995–1999 period at just under $20.7 billion, while the OSD Comptroller recently estimated the total at $18.2 billion; the difference is attributable to the inclusion or exclusion of various smaller operations. See Nina Serafino, *Peacekeeping: Issues of U.S. Military Involvement*, Washington, D.C.: Congressional Research Service Issue Brief, July 2000.

[56] The General Accounting Office has documented over the last decade a wide range of estimating errors in the FYDP. For example, the DoD has consistently underestimated the costs of base closings and environmental cleanup, depot and real property maintenance, military construction, medical care, and major weapon programs while overestimating the savings from personnel and infrastructure reductions and from defense reform.

[57] For example, $6.9 billion was added to the FY 1996 National Defense Authorization, $9.6 billion to FY 1997, and $3.8 billion to FY 1998.

[58] The Department of Defense—Military account benefited from an estimated $25.2 billion in emergency supplementals over the FY 1995–1999 period. See Congressional Budget Office, *Emergency Spending Under the Budget Enforcement Act: An Update*, Washington, D.C., June 8, 1999.

plans were revised upward so that the next budget request would better approximate the actual costs. Thus, there seems to have been tacit agreement between the executive and legislative branches that the caps on discretionary defense spending were sacrosanct and not subject to further debate—and that the annual defense appropriation process and emergency supplementals would be used in combination to address at least some of the recurring shortfalls. This resulted in a fair amount of churning while generally failing to close the gap.

In addition, the mismatch resulted in a focus on the short term at the expense of longer-term considerations. The postponement of spending on modernization and recapitalization resulted in both increased O&S costs as the costs of maintaining older systems rose and, by 1997, a renewed threat of precisely the sort of future procurement bow wave that the BUR had sought to avoid.

The View from the Air Force. The actual execution of the BUR's defense program and budget was generally in opposition to the strategic choice the Air Force had made to trade force structure for modernization; the Air Force's ability to pursue modernization was severely constrained by available budgets.

By FY 1998, the Air Force's budget authority had declined by nearly one-third (32 percent) since FY 1990, with roughly 22 percent of that decline in the Base Force years, another 6 percent during the FY 1994 transition year, and another 4 percent during FY 1995–1998. While spending fell somewhat unevenly across various Air Force major force programs, accounts, and titles, modernization was in general the principal source of savings. The cumulative decline in spending by Air Force Major Force Program from FY 1990 to FY 1998 varied by type of force: 72 percent for strategic forces; 34 percent for general-purpose forces; and 71 percent for special operations forces. Meanwhile, spending on airlift increased by 32 percent over the same period, while spending on the guard and reserve did not change.

While the Air Force was willing to trade additional force structure for modernization, ironically, spending on investment accounts declined. Investment spending fell from an estimated 43.9 percent in FY 1993 (the last Base Force budget) to 38.2 percent in FY 1995 and then climbed to 40.4 percent in FY 1998, the last BUR budget. Procurement accounts were the hardest hit, falling from 27 percent in budget authority in FY 1993 to 19 percent in FY 1997 before nudging up to 20 percent in FY 1998. This reduction in planned investment spending had a dramatic impact on Air Force procurement of aircraft. Further, spending on aircraft procurement for the Air Force generally fell below planned budget authority and outlays over the period.[59] As a result, the number of aircraft acquired by the Air Force declined dramatically after the FY 1991 budget and program and never recovered.[60] In addition, actual Air Force spending on RDT&E generally fell below planned spending as well.

ASSESSMENT

Capability to Execute the Strategy

The BUR established the following standard for evaluating its ability to execute its strategy:

> To achieve decisive victory in two nearly simultaneous major regional conflicts and to conduct combat operations characterized by rapid response and a high probability of success, while minimizing the risk of significant American casualties.[61]

As a practical matter, the overwhelming military capabilities of the United States in relation to other actors at the time (especially the greatly reduced Iraq) left little doubt that the United States would ultimately prevail in such conflicts. The issue was thus the degree of risk that the program was incurring—e.g., whether the capabilities were sufficient to defeat the enemy as quickly as desired, how much of the force would need to be engaged in both conflicts, and whether casualties could be minimized.

---

[59] In three out of the four years of FY 1995–1998, the difference between planned and actual budget authority for that year was more than $225 million; actual outlays fell below planned outlays in FY 1995 and FY 1996 only.

[60] Important exceptions to this trend were the acquisition of trainers, including the Tanker and Transport Trainer System and the Joint Primary Aircraft Trainer System (JPATS).

[61] Aspin, *Report on the Bottom-Up Review*, p. 8.

The 1993 JMNA reported that the forces programmed in the FY 1994 President's Budget request were adequate to achieve national security objectives with low to moderate risk—a risk level comparable to that assessed for the Base Force in 1993. Two main areas of concern, however, were identified by the 1993 JMNA. First, while conventional force capabilities were deemed adequate, continued deficiencies were found in rapid strategic lift, supporting elements, and sustainment. The JMNA judged that the readiness of the forces at the time made them capable of executing the two-MRC strategy but that mobility assets were insufficient to provide an acceptable level of risk.[62] In testimony on the BUR, representatives of the DoD indicated that when compared with the Base Force, the BUR force incurred a higher level of risk in executing the two-MRC strategy even with the planned force enhancements.[63] By July 1999, after the successful conclusion of the air war over Serbia, the Joint Chiefs were assessing the risk associated with the two-MRC strategy as high, with most of these risks tied to the ability to conduct operations on a second front.[64]

Readiness

The data suggest that military readiness trends through 1993 were generally quite favorable,[65] and in FY 1994 some of the most noteworthy readiness problems were in fact related to the disestablishment of units.[66] Other reviews from the period suggest that readi-

---

[62] Joint Chiefs of Staff, *Joint Military Net Assessment*, 1993, p. 3.

[63] See the questions and answers following DoD and Joint Staff testimony in U.S. Congress, Senate, "Force Structure Levels in the Bottom-Up Review," pp. 687–753.

[64] See Department of Defense, *Quarterly Readiness Report to the Congress, April–June 1999*, Washington, D.C., July 1999, p. 3.

[65] See Congressional Budget Office, *Trends in Selected Indicators of Military Readiness, 1980 Through 1993*, Washington, D.C., March 1994.

[66] For example, two of three late-deploying Army divisions that had experienced readiness problems in the previous year had fallen from C-2 to C-3 as a consequence of their planned disestablishment. See Chairman Shalikashvili's testimony before the Senate Armed Services Committee hearings on August 4, 1994.

ness levels were generally stable and consistent with service goals from 1990 through March 1996.[67]

However, a June 1994 Defense Science Board study of readiness noted the existence of "pockets" of unreadiness.[68] By February 1998, the services were reporting a variety of readiness problems, including inadequate funding for Army operations, training, and modernization;[69] increased strain on Air Force personnel because of high operational tempos, aging aircraft, and the need to rotate deployed forces throughout several forward-deployed locations;[70] lower levels of readiness among naval forces;[71] and aging equipment in the Marine Corps.[72]

By this time, the readiness issue had turned into a full-out debate.[73] Disagreement focused on whether the anecdotal evidence of readi-

---

[67] See U.S. General Accounting Office, *Military Readiness: Data and Trends for January 1990 to March 1995*, Washington, D.C., GAO/NSIAD-96-111BR, March 1996, and U.S. General Accounting Office, *Military Readiness: Data and Trends for April 1995 to March 1996*, Washington, D.C., GAO/NSIAD-96-194, August 1996.

[68] See John Deutch, "Memorandum for Distribution, Subject: Final Report of the Defense Science Board Task Force on Readiness," July 1994. See also Office of the Assistant Secretary of Defense (Public Affairs), "Readiness Task Force Presents Its Findings," OASD(PA) News Release No. 437-94, July 22, 1994, and Defense Science Board, *Report of the Defense Science Board Task Force on Readiness*, June 1994.

[69] Without a timely nonoffset supplemental, the Army argued that "there will be a devastating effect on Army readiness. The specific impacts include the decline of divisions to C-3 readiness levels with some likely to drop to C-4, cancellation of all remaining Combat Training Center rotations, cancellation of Army participation in remaining Joint Exercises, and elimination of virtually all collective Home Station training." See Lieutenant General Thomas N. Burnette, Jr., Deputy Chief of Staff for Operations and Plans, U.S. Army, testimony before the House Armed Services Committee, Subcommittee on Military Readiness, March 18, 1998.

[70] See Lieutenant General Patrick K. Gamble, Deputy Chief of Staff, Air and Space Operations, U.S. Air Force, testimony before the House Armed Services Committee, Subcommittee on Military Readiness, March 18, 1998.

[71] See Vice Admiral James O. Ellis, Jr., Deputy Chief of Naval Operations (Plans, Policy, and Operations), testimony before the House Armed Services Committee, Subcommittee on Military Readiness, March 18, 1998.

[72] See Lieutenant General Martin R. Steele, Deputy Chief of Staff for Plans, Policy, and Operations, testimony before the House Armed Services Committee, Subcommittee on Military Readiness, March 18, 1998. Lieutenant General Steele also detailed a number of coping mechanisms the Marine Corps was using to maintain readiness.

[73] In addition to congressional testimony on the readiness issue, see Floyd D. Spence, "Statement of Honorable Floyd D. Spence, Fiscal Year 1998 SECDEF/CJCS Posture

ness shortfalls was supported by more systematic measurement and on the actual nature of the short- and long-term effects on readiness of what all acknowledged were high levels of U.S. participation in peace operations. In the fall of 1998, the final FY 1998 quarterly readiness report to Congress described generally declining readiness trends for both combat and support forces, citing such deficiencies as resource shortfalls, aging and wearing equipment, and training shortfalls.[74]

A review of available evidence suggests that concern about readiness increased over the FY 1994–1998 period and that such concern was warranted. While not all forces were experiencing readiness problems—and while some of these problems were related to cyclical deployment schedules and budget calendars—such problems were neither isolated nor abating.

Modernization

By 1997, it had become clear that high rates of deployment and tempos of operations were eroding not only readiness and the capability of the force to execute the national military strategy but also the longer-term modernization and recapitalization effort (see Figure 3.8). As Table 3.6 and Figure 3.8 suggest, over the 1995–1997 period, spending on modernization remained well below the level planned in the FY 1994 and 1995 (BUR) budgets; funds routinely "migrated" from investment accounts to O&S accounts, resulting in program stretch-outs and delays to planned modernization efforts. In the

---

Hearing," February 12, 1996, and *Military Readiness 1997: Rhetoric and Reality*, House Committee on National Security, April 9, 1997; Ryan, *Military Readiness, Operations Tempo (OPTEMPO) and Personnel Tempo (PERSTEMPO)*; Dov Zakheim, "Global Peacekeeping Burden Strains U.S. Capability," *Defense News*, April 6, 1998, p. 19; Gordon Adams, "Contingencies Serve Role," *Defense News*, April 13, 1998, p. 21; John McCain, "Status of Operational Readiness of U.S. Military Forces," *Congressional Record*, Senate, September 10, 1998, pp. S10198–S10201; and John McCain, "Defense Preparedness," *Congressional Record*, Senate, September 30, 1998, pp. S11139–S11142. Zakheim and Adams were advisers to the 2000 Bush and Gore campaigns, respectively.

[74]Of the readiness deficiencies identified, approximately 70 percent were "capability" related, reflecting a lack of resources to meet established mission requirements, while 30 percent were due to "readiness" deficiencies that reflected a degradation in ability attributable to shortfalls in equipment condition or training.

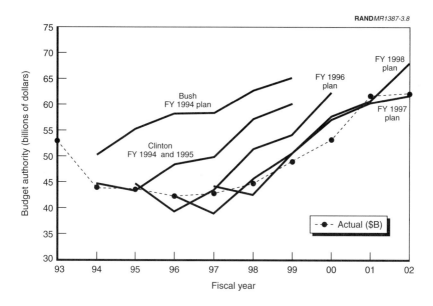

Figure 3.8—Planned and Actual Procurement, FY 1994–1998 Plans

end, the result was the creation of precisely the sort of bow wave that the BUR had planned to redress.

SECTION CONCLUSIONS

The BUR's strategy of engagement, prevention, and partnership laid the groundwork both for the national security strategy of engagement and enlargement that was to follow and for the national military strategy that posited two nearly simultaneous MRCs as the basis for force sizing and, ultimately, for assessing readiness and strategic risk. Many or most of the force structure goals were achieved by FY 1996–1998, with some accomplished as early as FY 1994, before implementation of the BUR had formally begun.[75] However, many of the force enhancements required to make the strategy work with the reduced force structure—for example, in the area of strategic mobil-

---

[75]These included reserve-component Army divisions, Air Force TFWs, Navy aircraft carriers, and Marine end strength.

ity—were not in place by 1999 as had been expected. The result of these multiple shortfalls appears to have been a higher level of risk in executing the military strategy at the end of the BUR period than had been anticipated. Moreover, although prior budgetary guidance greatly constrained both strategy and force structure, it seems to have done little to limit the employment of the U.S. military over the period.

It is important to note the existence of several important "disconnects" in the BUR. Most significantly, despite emerging indicators that this might be appropriate, the BUR did not reexamine in its consideration of "conflict dynamics" the BUR's assumption that peace operations could be treated as "lesser-included cases" that would impose few costs and risks on readiness or warfighting capability.[76] Research in fact suggests that the cumulative level of peacetime operations approximated a full MRC or more of force structure.[77] As a result of the accumulation over time of large and/or long-duration commitments—and despite the readiness-monitoring panels that the BUR had endorsed—readiness problems and risks to warfighting capabilities increased over the period, the prospects for which had been underestimated by the BUR.

In retrospect, in lieu of option three—the strategy and force structure capable of two nearly simultaneous MRCs—option four, which would also have supported SSCs, might have provided a more substantial rotation base for contingency operations while mitigating the effects of high deployment tempos.[78] With the most capable strategy/force structure option (four) having been ruled out for reasons of cost, however, the result was an ambitious strategy

---

[76]This assumption appears to have been warranted in the case of the Base Force, however, precisely because policymakers did not as a matter of policy promote a substantial U.S. military commitment to peace operations.

[77]See Ryan, *Military Readiness, Operations Tempo (OPTEMPO) and Personnel Tempo (PERSTEMPO)*, pp. 10 and 13. According to former Deputy Under Secretary of Defense Louis C. Finch, the requirements for peacekeeping/peace enforcement and humanitarian operations were, in retrospect, understated.

[78]It may also be the case that if all the "critical enhancements" that aimed to improve strategic mobility or increase the density of what later came to be called low-density/high-demand (LD/HD) assets had been put in place, force structure might have been sufficient to underwrite the strategy.

supported by a reduced and ultimately underfunded force structure.[79]

In the end, the history of the BUR suggests the importance of reevaluating key assumptions of prior strategies and, when necessary, revising these assumptions and making changes to strategy, forces, or resources. The BUR also demonstrates, however, that coping mechanisms that fail to address the underlying balance between strategy, force structure, and resources may be limited in their ability to redress fundamental mismatches and that the failure to ask hard questions and establish clear priorities—between warfighting and peacetime operations, for example, or between investments in short- and long-term readiness—can ultimately lead to precisely the outcome that planners most seek to avoid: an increase in the risks associated with execution of the strategy, coupled with erosion in both short- and long-term force readiness.

The next chapter assesses the 1997 QDR and describes how that review sought to reestablish a better balance between strategy, forces, and resources and to extend the time horizon for defense planning beyond paying bills for current operations.

---

[79] Of course, with the budget fixed, option four would have been even more unaffordable.

Chapter Four
THE 1997 QUADRENNIAL DEFENSE REVIEW: SEEKING TO RESTORE BALANCE

The 1997 *Report of the Quadrennial Defense Review* was intended to provide a blueprint for a strategy-based, balanced, and affordable defense program.[1] Within a fixed budget of roughly $250 billion a year and with only modest adjustments to force structure, the QDR aimed to rebalance the defense program and budget to address one of the key problems that had developed during the BUR years: the "migration" of funds from modernization (and particularly procurement) accounts to operations accounts.

Three key factors militated against a successful outcome. First, CJCS John Shalikashvili seized the initiative in getting out in front of civilian policymakers in OSD while constraining the range of potential options by effectively ruling out any major changes to the status quo

---

[1] The QDR is documented in Department of Defense, *Report of the Quadrennial Defense Review*; Armed Forces Information Service, *Defense 97: Commitment to Readiness*, Washington, D.C., 1997; and Office of the Assistant Secretary of Defense (Public Affairs), "Quadrennial Defense Review," DoD news briefing, Washington, D.C., May 19, 1997, available at http://www.defenselink.mil/news/May1997/t051997_t0519qdr.html (accessed September 2000). For analyses of the 1997 QDR, see John Schrader, Leslie Lewis, and Roger Allen Brown, *Quadrennial Defense Review (QDR) Analysis: A Retrospective Look at Joint Staff Participation*, Santa Monica: RAND, DB-236-JS, 1999; James S. Thomason, Paul H. Richanbach, Sharon M. Fiore, and Deborah P. Christie, *Quadrennial Review Process: Lessons Learned from the 1997 Review and Options for the Future*, Alexandria, VA: Institute for Defense Analyses, IDA Paper P-3402, August 1998; U.S. General Accounting Office, *Quadrennial Defense Review: Opportunities to Improve the Next Review*, Washington, D.C., GAO/NSIAD-98-155, 1998; and Booz-Allen and Hamilton, *An Assessment of the National Review Process*, McLean, VA, December 1999.

regarding force structure, roles and missions, and service budget shares.[2] Second, Defense Secretary William Cohen's status as the newest member of (and sole Republican in) the Clinton cabinet gave him little of the compensating leverage that would be needed to overturn the status quo that Chairman Shalikashvili and the White House supported; it is not clear how much Secretary Cohen might have changed the result even if he had spent all of his political capital to that end. Third, budgets were effectively frozen at levels that, as described in Chapter Three, not only seemed incapable of resolving the emerging gaps but also hindered the development of more creative strategies for resolving the DoD's dilemma.[3]

Taken together, it seems to have been almost a foregone conclusion that the QDR would fail to achieve the balance that it ostensibly sought.

In the end, the QDR yielded relatively modest reductions to force structure and end strength in its attempts to stabilize the level of defense resources and better achieve modernization goals. Force structure reductions would be selective and minimal, with the resulting force structure looking for the most part like that which had resulted from the BUR. Active manpower would fall by 6.2 percent, reserve manpower by 7.2 percent, and civilian manpower by 20 percent below their 1997 levels. Budgets would remain roughly at the 1997 level of $250 billion, while procurement spending would increase to $60 billion by FY 2001. This chapter describes the balancing of strategy, forces, and resources in the design, planning, and execution of the QDR.

BUILDING THE QDR FORCE

Background

World Situation. The QDR described the current world situation as one of "strategic opportunity" for the United States: The threat of

---

[2]See George C. Wilson, *This War Really Matters: Inside the Fight for Defense Dollars*, Washington, D.C.: Congressional Quarterly Press, 1999, pp. 38–45.

[3]The QDR also assumed, somewhat unrealistically, that additional BRAC rounds would be approved and that additional infrastructure savings could therefore be assumed to be available during the execution of the QDR.

global war had receded, and U.S. core values of representative democracy and market economics had been embraced in many parts of the world. This situation was seen as having created new opportunities to promote peace, prosperity, and enhanced cooperation among nations. U.S. alliances, including NATO as well as those with Japan and Korea, were adapting successfully to meet the emerging challenges and providing "the foundation for a remarkably stable and prosperous world." Former adversaries like Russia and other former members of the Warsaw Pact were cooperating with the United States across a range of security issues. The result was that the United States was seen by many as the security partner of choice, the "sole superpower," and "the indispensable partner,"[4] and the QDR assumed that the conditions prevalent in the current security environment would continue to at least 2015. Within this time frame, no regional power or coalition was expected to amass sufficient conventional military strength to be able to defeat U.S. armed forces once the full military potential of the United States was mobilized and deployed.

At the same time, the QDR emphasized the emergence of "new threats and dangers—harder to define and more difficult to track."[5] Among these threats were regional dangers, including the threat of coercion and cross-border aggression in Southwest Asia, the Middle East, and East Asia; the proliferation of advanced technologies, including WMD; transnational dangers such as the spread of illegal drugs, organized crime, terrorism, and uncontrolled refugee migration; and threats to the U.S. homeland through terrorism, cyber attacks on computer networks, intercontinental ballistic missiles, and WMD.[6] Terrorist incidents during the period—both at home and abroad—heightened security concerns about forward-deployed forces and homeland defense.[7] One major reason for such concerns

---

[4]Department of Defense, *Report of the Quadrennial Defense Review*, p. 5.

[5]Op. cit., p. iii.

[6]Op. cit., p. 4.

[7]Among the most devastating and graphic of these incidents were Aum Shinrikyo's sarin attack on the Tokyo subway in March 1995, the bombing of the Alfred P. Murrah Building in Oklahoma City in April 1995, and the truck bombing of Khobar Towers in Saudi Arabia in June 1996. The bombing of the World Trade Center took place during the BUR study, on February 27, 1993. The Khobar Towers bombing led to Operation Desert Focus, the redeployment of U.S. forces to safer bases, in July 1996.

was that the unsurpassed military capabilities of the United States in conventional warfighting were expected to lead adversaries to develop strategies and capabilities for "asymmetric" attacks, both in theater and potentially against the U.S. homeland. Capabilities that rested on information technologies were seen as particularly vulnerable, including space-based, C^4, and intelligence, surveillance, and reconnaissance (ISR) assets.[8]

As shown in Chapter Three, these "new threats and dangers," coupled with the administration's activist conception of engagement and the military's role in promoting engagement, led to U.S. participation in a large number of military operations between the BUR and the QDR.[9] As was described, the average number of USAF aircraft deployed to contingency operations continued to increase over the period, from roughly 225 in October 1993 to more than 250 by June 1996.

Defense Resource Constraints. In 1997, the federal budget deficit continued to be a source of concern. Most worrisome were long-term trends that threatened to create what the Congressional Budget Office referred to as "unprecedented deficits and debt by the middle of the next century, potentially causing damage to the economy."[10] For planning purposes, the QDR therefore assumed a relatively sta-

---

[8]Ibid.

[9]The DoD reported more than 25 U.S. operations involving 500 or more personnel conducted between October 1993, when the BUR was published, and May 1997, when the QDR was published. See Department of Defense, *Report to Congress on U.S. Military Involvement in Major Smaller Scale Contingencies Since the Persian Gulf War*, Washington, D.C., March 1999, pp. 8–9. Vick et al. reported more than 85 Air Force military operations other than war (MOOTW) begun between October 1993 and December 1996. See Alan Vick et al., *Preparing the U.S. Air Force for Military Operations Other Than War*, Santa Monica: RAND, MR-842-AF, 1997.

[10]See Congressional Budget Office, *Reducing the Deficit: Revenue and Spending Options*, Washington, D.C., March 1997. By September 1997, the Congressional Budget Office was projecting a surplus by 2002 and saw few opportunities for further cuts in defense: "The peace dividend resulting from the end of the Cold War has probably been used up, and it is not clear that policymakers will agree to cut defense spending as they did in the mid-1990s. If anything, unforeseen conflicts elsewhere in the world and replacement of aging equipment could push defense spending in the opposite direction." Defense spending in 1997 was also expected to be $3 billion higher as a result of faster spending on O&M, procurement, and research and development. See Congressional Budget Office, *The Economic and Budget Outlook: An Update*, Washington, D.C., September 1997.

ble DoD budget of approximately $250 billion per year in constant FY 1997 dollars over the period of the assessment—roughly the budget described in the FY 1998 President's Budget and the FY 1998–2003 FYDP (see Table 4.1).[11] As had been the case with the BUR, the strategy and force options available to the authors of the QDR were thus to be greatly constrained by the resources that were assumed to be available.

Strategy Under the QDR

In developing a defense strategy, the QDR assumed that the United States would remain politically and militarily engaged in the world over the next 15 to 20 years and that it would maintain military superiority over current and potential rivals as the world's only superpower over the 1997–2015 period.[12]

Table 4.1

FY 1998 Long-Range Forecast for DoD Spending

	1997	1998	1999	2000	2001	2002	2003
Then-year $B							
BA	250.0	250.7	256.3	262.8	269.6	277.5	284.6
Outlays	254.3	247.5	249.3	255.2	256.2	261.4	276.1
FY 1997 $B							
BA	250.0	244.3	244.4	244.9	245.4	246.7	246.9
Outlays	254.3	241.2	237.7	237.9	233.4	232.7	240.3

SOURCE: DoD Comptroller, *National Defense Budget Estimates, FY 1998*, Washington, D.C., March 1997.

---

[11] Department of Defense, *Report of the Quadrennial Defense Review*, p. 19. As shown in Table 4.1, this plan in fact anticipated modest real reductions from FY 1997 spending levels. The Congressional Budget Office reported in January 1997 that approximately $273 billion would be required in FY 1998 to preserve real defense spending at 1997 levels. See Congressional Budget Office, *The Economic and Budget Outlook: Fiscal Years 1998–2007*, Washington, D.C., January 1997, p. 32. Defense Secretary Cohen claimed that when he entered office, he was told that the budget was "fixed by the highest number between the Executive Branch and the Congress, and that would be it for the foreseeable future." See "Summing Up: Cohen," *The NewsHour with Jim Lehrer*, Public Broadcasting System, January 10, 2001. For some thoughts on the implications of the politics of surplus for defense spending, see the appendix.

[12] Department of Defense, *Report of the Quadrennial Defense Review*, p. 5.

In establishing its defense strategy, the QDR set up two major decisions, each characterized by a choice among two highly stylized (and unattractive) straw-man "options" and a more moderate (and infinitely more attractive) middle path.

First, a decision was made regarding the nature of the United States' role in the world. The QDR rejected a form of isolationism that argued that "our obligations beyond protecting our own survival and that of key allies are few."[13] It also rejected what might be called acting as a "world policeman," where the United States would take on "obligations that go well beyond any traditional view of national interest, such as generally protecting peace and stability around the globe, relieving human suffering wherever it exists, and promoting a better way of life, not only for our own citizens but for others as well." In lieu of these extreme views, the QDR advocated a more balanced course of "engagement" that presumed that the United States would continue to exercise strong leadership in the international community using all dimensions of its influence to shape the international security environment. This approach was viewed as particularly critical to ensuring peace and stability in regions where the United States had vital or important interests and to broadening the community of free-market democracies.

Second, the QDR made a decision regarding how best to allocate resources among three elements of a new strategy: *shaping* the international security environment in ways favorable to U.S. interests, *responding* to the full spectrum of crises when directed, and *preparing now* for an uncertain future by transforming U.S. combat capabilities and support structures to shape and respond effectively to future challenges.

This choice also involved charting a middle course between unattractive extremes. In this case, the QDR rejected a defense posture focused solely on near-term demands as well as one focused chiefly on distant threats in favor of a posture that balanced current demands against the needs of an uncertain future. The objective of the chosen posture was to meet both near- and long-term challenges, taking the position that the world did not afford the United

---

[13]Op. cit., pp. 7–8.

States the opportunity to choose between the two. While this strategy had much in common with the BUR's strategy of engagement and enlargement, the third element of the strategy—preparing now for an uncertain future—sought to address one of the key problems that had emerged during the BUR years: inadequate and unreliable funding for the modernization of the force.

Assumptions About Future Operations. The QDR argued for the necessity of maintaining the United States' unparalleled military capabilities and of assuring both its continued status as the predominant military power and its preparedness to undertake the wide range of missions anticipated over the 1997–2015 period.[14]

The capability to fight and win two MTWs was seen as the high end of the crisis continuum, the most stressing requirement for the U.S. military, and the hedge against even more difficult contingencies;[15] the QDR announced that its force needed to be capable of executing two nearly simultaneous MTWs with moderate risk. As in the BUR, the two-war capability was dictated primarily by the potentially adverse consequences of having only a one-war capability; as stated in the QDR, "If the United States were to forgo its ability to defeat aggression in more than one theater at a time, our standing as a global power, as the security partner of choice, and as the leader of the international community would be called into question."[16] In comparison with the BUR, the QDR placed greater emphasis on the quick-halt phase in MTWs, arguing that the United States needed to be able to rapidly defeat initial enemy advances short of their objectives in two theaters in close succession, one followed almost immediately by another.

U.S. forces could also be expected to participate in a great many SSCs involving a broad array of operations: show-of-force operations, interventions, limited strikes, noncombatant evacuation operations, no-fly-zone enforcement, peace enforcement, maritime sanctions enforcement, counterterrorism, peacekeeping, humanitarian assis-

---

[14]Op. cit., p. 8.
[15]Op. cit, p. 12.
[16]Ibid.

tance, and disaster relief.[17] Indeed, the demand for SSCs was expected to remain high over the next 15 to 20 years and to pose the most frequent challenge through 2015.

Although the QDR gave increased rhetorical attention to the demands of SSCs, no fundamental change was made to the BUR's basic approach, which stressed the necessity of managing "conflict dynamics"—preparations to allow for a quick transition from a posture of peacetime engagement to warfighting—and relied on management review to minimize SSC-related deployment and personnel tempos, readiness, and other risks to warfighting capabilities.[18] The QDR argued that U.S. forces would need to ensure the capability to transition from global peacetime engagement or multiple concurrent SSCs to fighting MTWs[19]—a particularly important capability in light of the QDR's recognition of the growing potential for multiple concurrent SSCs and regional engagement missions. The QDR also gave emphasis to new missions, including counterproliferation, force protection, counterterrorism, and information operations.

In defining the parameters of future U.S. uses of force, the QDR articulated a somewhat more cautious and nuanced employment doctrine than had the BUR, distinguishing among situations involving "vital," "important but not vital," and "humanitarian" interests and identifying the sorts of responses appropriate to each.[20] Like both the BUR and the Base Force, the QDR reported that it favored a coalition strategy that, while preserving the United States' ability to act unilaterally, emphasized cooperative, multinational approaches—including coalition operations that would enhance political legitimacy and distribute the burden of responsibility among like-

---

[17] Op. cit., p. 11.

[18] For example, the QDR endorsed the Global Military Force Policy, which allocates LD/HD assets across competing priorities. Op. cit., p. 36.

[19] The QDR noted: "In the event of one major theater war, the United States would need to be extremely selective in making any additional commitments to either engagement activities or smaller-scale contingency operations. We would likely also choose to begin disengaging from those activities and operations not deemed to involve vital U.S. interests in order to better posture our forces to deter the possible outbreak of a second war. In the event of two such conflicts, U.S. forces would be withdrawn from peacetime engagement activities and smaller-scale contingency operations as quickly as possible to be readied for war." Op. cit., p. 13.

[20] Op. cit., p. 9.

minded states, particularly among the world's most influential countries.[21]

The View from the Air Force. Significantly, the QDR's increased emphasis on the halt phase in large-scale cross-border aggression placed an even greater premium on quickly deployable air-to-ground strike capabilities and strategic mobility than had the BUR four years earlier. Put another way, not only had the halt-buildup-counteroffensive construct that had been introduced in the BUR become part of the canon, but that canon increasingly favored aerospace forces.

That said, Air Force Chief of Staff Ronald Fogleman appears to have been even more disappointed by the QDR's failure to address future roles and missions and the restructuring of the force than General McPeak had been by the BUR's.

General Fogleman had entered the QDR with high expectations for what it might hope to accomplish:

> Viewing the Air Force from the outside as military historian, as someone who has tried to stay involved in academic affairs as well as national security affairs—I sincerely believed that the nation was at a unique crossroads, that the country had a tremendous number of internal needs, that the external threats were lower than we had faced in half a century, and that we had an opportunity—if we could have a serious discussion about national security strategy and defense issues—to restructure our military into a smaller, better focused institution to respond to the kinds of challenges coming in the next 10 to 15 years. It was not a military that was going to be shaped by some force-structure slogan like two MRCs, and it had to include a fundamental understanding of whether there really was a "revolution in military affairs" and how we could and should fight future wars. So I had begun to speak out about the Quadrennial Defense Review, and I was hopeful that the QDR would start us down that path. In this regard, in "the tank" I began to question some of the things that we were doing, or that we were planning to do, based on old paradigms—but not very successfully.[22]

---

[21] Op. cit., p. 8.

[22] Richard H. Kohn, "The Early Retirement of General Ronald R. Fogleman, Chief of Staff, United States Air Force," *Aerospace Power Journal*, Spring 2001, pp. 11–12.

These ambitions were to remain unrealized, however, as a result of additional constraints that were imposed on the ambit of the QDR by Chairman Shalikashvili in the early fall of 1996:

> As we began talking more and more about the QDR, an event occurred in September 1996 which kind of put the QDR in a context that struck me as all wrong. An Army two-star from the JCS came by to see all the chiefs, and when he came to see me, he sat on that couch in the chief's office and said, "I have a message from the chairman, and the message is, that in the QDR we want to work hard to try and maintain as close to the status quo as we can. In fact, the chairman says we don't need any Billy Mitchells during this process." That shocked me a little bit. I replied, "Well, that's an unfortunate use of a term, but I understand the message." From that point on, I really did not have much hope for the QDR.[23]

> I guess I lost all hope when Bill Perry left because he had the stature to have given the services the blueprint, and I think the services would have fallen in line. . . . Once Bill Perry left, work on the QDR went into suspended animation until Cohen arrived because no one wanted to get out in front of the new boss. He arrived with a very limited amount of time to deliver the QDR to the Hill, a difficult challenge. I came to believe that the QDR could not be completed in three months, or even six. To an extent, he tried to solicit the advice of his military people, but it became clear that this QDR was to be more a political response than a sincere effort to reshape our military. It was driven by the consideration to come up with $60 billion in savings to apply to the procurement of new weapons.[24]

General Fogleman was particularly critical of the process by which the decision was made in the QDR to further reduce key theater air modernization programs, including the Air Force's own F-22.[25]

---

[23] Op. cit., pp. 12–13. General Fogleman has also stated that after receiving Chairman Shalikashvili's message, he went from expecting that something good might come out of the QDR to working to minimize the damage that might otherwise result from program cuts. Op. cit., pp. 40–42.

[24] Ibid.

[25] Op. cit., pp. 13–14.

Building the Force

As a result of changes in both the nature of the threats faced and the way in which future conflicts would be fought, the forces and capabilities required to uphold the QDR's two-theater strategy differed somewhat from the MRC building blocks used in the BUR.[26] As was described above, the QDR gave somewhat greater rhetorical emphasis than had the BUR to the force requirements of other contingencies:

> While the Bottom-Up Review focused primarily on that difficult task [two nearly-simultaneous MTWs], we have also carefully evaluated other factors, including placing greater emphasis on the continuing need to maintain continuous overseas presence in order to shape the international environment and to be better able to respond to a variety of smaller scale contingencies and asymmetric threats.[27]

In shaping its forces, the QDR also embraced Joint Vision 2010, the capstone vision for future joint warfighting that had identified "full-spectrum dominance" as the preeminent goal for planning future military capabilities. Among the desired characteristics of the full-spectrum force described in the QDR were flexibility and versatility to succeed in a broad range of anticipated missions and operational environments; sufficient size and capability to defeat large enemy conventional forces, deter aggression and coercion, and conduct the full range of SSCs and shaping activities, all in the face of asymmetric challenges; and multimission capability, proficiency in warfighting competencies, and the ability to transition from peacetime activities and operations to enhanced deterrence in crises, and then to war.

Assignment of Forces for Overseas Presence. The deployment of troops overseas aimed to stress continuity and to signal the United States' commitment to peace and stability in Europe and in the Asia/Pacific region, with an anticipated 100,000 military personnel

---

[26]Op. cit., p. 13. The principal reason given was that the "accelerating incorporation of new technologies and operational concepts into the force calls for a reexamination of the forces and capabilities required for fighting and winning major theater wars."

[27]Op. cit., p. v.

Table 4.2

Assignment of Forces in the QDR, May 1997

Force Package	Army Divisions	USAF TFWs	USMC MEFs	Navy Carriers[a]
Forward				
Atlantic				
Europe	2.0	2.3		0.7
Pacific				
Japan	0.11[b]	1.4	1.0	1.0
South Korea	0.66[c]	1.0	(0.66)[d]	
Persian Gulf	—[e]	—[e]	(0.33)[f]	0.7–1.0[g]
Total	2.66	4.7	1.0	2.4–2.7
Contingency				
U.S.				
Active	7.33	8.1	2.0	7.3–7.6
Reserve	8.0	8.0	1.0	1.0[h]
Total	15.33	16.1	3.0	8.3–8.6
Total	18.0	20.8	4.0	12.0
Active	10.0	12.8	3.0	11.0
Reserve	8.0	8.0	1.0[i]	1.0[h]

SOURCES: Department of Defense, *Report of the Quadrennial Defense Review*, Washington, D.C., May 1997, and authors' estimates.

[a]Eleven active plus one reserve carrier capable of sustaining full-time presence in one region and presence in two other regions 70 percent of the time. The Navy never implemented the reserve carrier concept.

[b]Army Special Forces battalion in Okinawa, not scored against division count.

[c]The BUR plan called for ultimate reduction to one brigade.

[d]Two brigade-size MEFs (two MPS squadrons) available for the MTW in Northeast Asia.

[e]Land-based Army and Air Force forces to be rotational only.

[f]One MPS squadron for brigade-size MEF available for the MTW in Southwest Asia.

[g]"Tether" carrier, supplemented by other Navy ship and a rotational ARG.

[h]Operational reserve/Navy carrier. This carrier was subsequently returned to the active force.

[i]Reserve Marine division/wing/service support group.

assigned in each region. As shown in Table 4.2, the resulting assignment of forces differed only modestly from that of the BUR; the larger number of reserve Army divisions (from 15+ to 18) reflected the post-BUR summit decisions, and Air Force TFWs were modestly increased (from 20 to 20.8) and restructured (from 13.0 to 12.8 active wings and from 7 to 8 reserve wings). Rotational deployments of active and reserve naval, air, and ground forces to other key regions such as Southwest Asia were also to continue, as were planned improvements to afloat and ashore prepositioned stocks of equipment and materiel.

The resulting force structure and assignment of forces were judged by the QDR to be adequate to provide a balanced capability for meeting the full range of anticipated future demands.

Force Structure and Manpower. As noted above, the force options available to QDR policymakers were greatly constrained by the limited resources that would be available. The QDR concluded, however, that a 10 percent force structure cut would result in unacceptable risk in implementing the national military strategy. Accordingly, changes to force structure involved only modest reductions as well as some restructuring of forces.[28] Among the most important of these changes was the decision to move one Air Force TFW from the active to the reserve component, leaving slightly more than 20 TFWs in the force structure. Other critical changes included reductions in reserve air defense squadrons (from 10 to 4) and in Navy attack submarines (from 73 to 52) and surface combatants (from 131 to 116).

With more substantial force structure cuts off the table and with savings from further infrastructure reductions deemed insufficient, the QDR viewed reductions to manpower as the principal source of needed savings. The Secretary of Defense accordingly directed the services to develop plans to cut the equivalent of 150,000 active military personnel to provide $4 billion to $6 billion in recurring savings by FY 2003. The services responded with analyses that led to initiatives to eliminate some 175,000 personnel and save an estimated $3.7 billion. In the case of the Air Force, these cuts were to result in a reduction of 7 percent of active military personnel, 0.4 percent for the reserves, and 11 percent for civilian personnel by FY 2003.

The View from the Air Force. In addition to the shift in the active/reserve mix of fighter forces, the following decisions taken by the QDR also were important to the Air Force:

---

[28]Op. cit., p. 30. See also John A. Tirpak, "Projections from the QDR," *Air Force Magazine*, Vol. 80, No. 8, August 1997.

- *Manpower* was to be reduced by 26,900 active, 70 reserve, and 18,300 civilian personnel. The USAF chose to try to achieve many of these reductions through aggressive competitive outsourcing.

- The *tactical force* was to see restructuring and modest reductions to fighter inventory, including retiring old Air National Guard (ANG) fighters and replacing them with about 60 active fighters and converting six continental air defense squadrons to general-purpose squadrons.

- *Long-range bombers* were to be maintained at a total of 187, with 142 to be assigned to operational units.

- Many Air Force *high-leverage and specialized assets* (e.g., bombers, F-117s, standoff jammers, AWACS, JSTARS, and C^{4}ISR platforms) were to swing from the first major conflict to the second one.

- The *tanker and airlift fleet* was to see no major changes. In accordance with the Mobility Requirements Study Bottom-Up Review Update (MRS BURU), total airlift capacity was to be sized at approximately 50 MTM/D.

- Finally, $64 million was to be added to *strategic forces* in FY 1999 to maintain START I levels until the Russian Duma ratified START II.

Resources

The QDR acknowledged that the principal failing of the BUR's defense spending plan was that it had underestimated operating costs and overestimated savings and that the resulting offsets had reduced resources for modernization as they "migrated" to operations accounts (see Figure 4.1). Accordingly, the QDR sought to reallocate "resources and priorities to achieve the best balance of capabilities for shaping, responding, and preparing over the full period covered by the Review." Equally important, however, was providing a budget and program that were fiscally responsible and capable of successful execution:

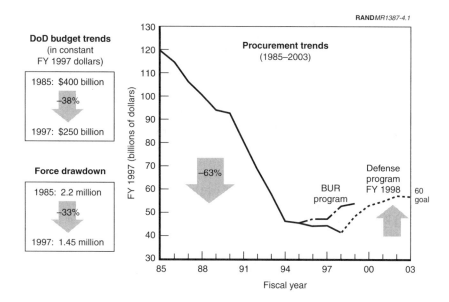

Figure 4.1—The QDR's Investment Challenge

The direct implication of this fiscal reality is that Congress and the American people expect the Department to implement our defense program within a constrained resource environment. The fiscal reality did not drive the defense strategy we adopted, but it did affect our choices for its implementation and focused our attention on the need to reform our organization and methods of conducting business.[29]

As part of this allocation, and as suggested by the various force structure, manpower, modernization, infrastructure, and funding decisions, the QDR sought to trim both combat and support capabilities while preserving funding for the next generation of weapon systems.

To better support modernization, the QDR reported that the FY 1998 President's Budget and the six-year FYDP had projected an increase in procurement funding from $42.6 billion in the FY 1998 budget to

---

[29]Department of Defense, *Report of the Quadrennial Defense Review*, p. v.

$60 billion in nominal terms by FY 2001.[30] It also noted, however, that the spending plan had the continued potential for annual migration of funds to unplanned operations expenses of as much as $10 billion to $12 billion per year, creating the possibility that rather than achieving the $60 billion target, procurement funding might stall in the range of $45 billion to $50 billion. In addition, the QDR's stable budget of roughly $250 billion per year in constant FY 1997 dollars over the assessment period implied that any increases in procurement would need to be offset by other accounts or through additional savings.[31]

Defense Priorities. Rather than undertake a detailed analysis of modernization funding needs—either in terms of the funding needed for the overall recapitalization of the force[32] or for the transformation of the force—the QDR essentially accepted and embraced the $60 billion-a-year procurement spending goal advanced by Chairman Shalikashvili in 1995. In so doing, it in fact seems to have reduced procurement spending from the level established for FY 1998 (see the dashed line in Figure 4.2), thereby enhancing the prospects that spending targets actually could be achieved.

The aim of the QDR's selective modernization plan was to preserve funding for the next generation of systems, including information systems, strike systems, mobility forces, and missile defense systems, "that will ensure our domination of the battlespace in 2010 and beyond."[33] Accordingly, the QDR, like the BUR, advocated "selective" and "focused" modernization of U.S. weapon systems.

---

[30]This is roughly $54.9 billion in constant FY 1997 dollars, or $56.1 billion in constant FY 1998 dollars. The origins of the $60 billion target that was later cited by CJCS John Shalikashvili are to be found in the analyses supporting the Defense Program Projection. The figure reflected what was believed to be a minimum level of procurement and was originally calculated in terms of constant FY 1993 dollars. By the time of the QDR, the number was being treated in nominal rather than constant terms.

[31]Department of Defense, *Report of the Quadrennial Defense Review*, p. 19.

[32]For example, a SECDEF-level decision on the issue of the post-FYDP modernization bow wave was deferred while decisions were made on selective modernization issues such as tactical air, restructuring THAAD (the theater high-altitude area defense missile), accelerating national missile defense, and restructuring V-22 procurement. See Schrader, Lewis, and Brown, *Quadrennial Defense Review (QDR) Analysis: A Retrospective Look at Joint Staff Participation*, Appendix B, "QDR Issues and References."

[33]Department of Defense, *Report of the Quadrennial Defense Review*, p. v.

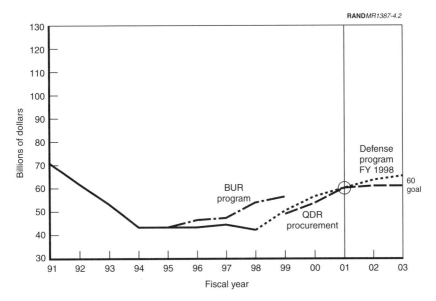

Figure 4.2—Planned Procurement Under the BUR, FY 1998 Defense Program, and QDR

At the same time, reductions in the quantities of many new weapon systems were seen as the source of potential savings. Thus, the JSTARS buy was reduced from 19 to 13, predicated on the expectation that the United States' NATO partners would purchase four to six Air Ground Surveillance (AGS) aircraft. The theater air program was subject to cuts, with procurement quantities reduced for the Air Force F-22 (from 438 to 339 aircraft), the Navy F/A-18E/F (a minimum of 548 aircraft), and the Joint Strike Fighter (from 2978 to 2852 aircraft). The decision was also made not to procure additional B-2s beyond the planned force of 21 aircraft.

Other programs, however, were supported at previously planned or even increased levels. The procurement of a tenth Nimitz-class aircraft carrier was authorized, for example, and a production rate of one and one-half to two attack submarines a year was supported. The acceleration of an Army "digitized division" was supported as well, as were the RAH-66 Comanche and Crusader self-propelled howitzer. Research and development on national missile defense was accelerated in anticipation of a deployment decision in FY 2000.

National cruise missile defense was also to receive increased emphasis, as were capabilities to counter the so-called asymmetric threats posed by nuclear, biological, and chemical weapons.

The QDR planned to fund modernization goals through infrastructure reduction and management reforms. Two additional rounds of BRAC were sought, and major initiatives were put forward to reengineer and reinvent DoD support functions with an increased emphasis on using the private sector to perform nonwarfighting support functions.[34]

IMPLEMENTING THE QDR

Strategy

In some respects, the new strategy elements of shaping and responding differed little from the BUR's strategy of engagement and enlargement or from the Base Force's reliance on forward presence and crisis response: All relied heavily on forward presence and crisis response capabilities, and all were concerned with ensuring stability in the near term in regions of vital interest.

However, the QDR resulted in much higher-than-expected levels of U.S. participation in peace and other contingency operations.[35] In February 1998, CJCS Henry Shelton reported that 1997 had seen 20 major operations and many smaller ones, with an average of 43,000 service members per month participating in contingency operations. As shown in Figure 4.3, Air Force participation in contingency operations over the period also remained quite high.

The largest ongoing Air Force commitments—and the greatest turbulence—continued to be associated with U.S. operations in Southwest Asia and the Balkans. The peak in February 1998 was attributable to Desert Thunder—the increase in U.S. forces in

---

[34]See U.S. General Accounting Office, *Defense Reform Initiative: Organization, Status, and Challenges*, Washington, D.C., GAO/NSIAD-99-87, April 1999, p. 15.

[35]See Jim Garamone, "Shelton Warns of Readiness Problems," American Forces Information Service, October 1, 1998.

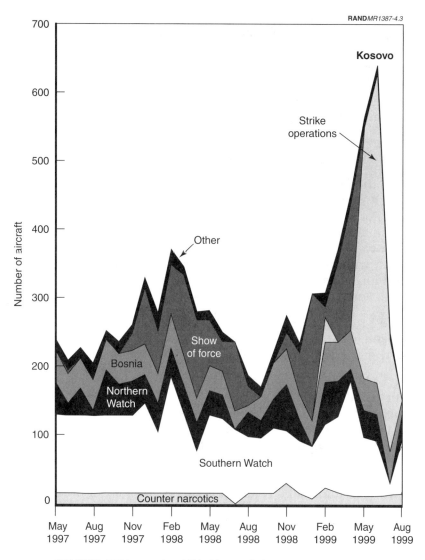

Figure 4.3—USAF Aircraft in Contingency Operations, 5/97–8/99

Southwest Asia in anticipation of strikes on Iraq[36]—while that in the spring of 1999 was attributable to the war in Kosovo, which was the largest use of Air Force combat forces since the 1991 Gulf War. These ongoing commitments and responses to periodic "pop-up" crises have ensured that the Air Force has remained at high operational and personnel tempos to the present day, mitigated somewhat by such innovations as the aerospace expeditionary force (AEF).

Force Structure and Manpower

Given the modest changes to force structure recommended by the QDR, it should come as little surprise that force structure changes for major force elements were, with only a few exceptions, already in place in the FY 2001 President's Budget and defense program.[37] The majority of QDR force structure changes were accomplished rather quickly with a few exceptions, including the additional reductions to naval forces and the eight-wing goal for the reserve component of the Air Force.

The FY 1999 President's Budget and defense program that was to implement the planned QDR manpower reductions was generally in line with the decisions and plans in the QDR (see Table 4.3).

The FY 2000/2001 President's Budget request and defense program anticipated that by the end of FY 2001, active forces would be at roughly 1.38 million, selected reserves at 866,000, and civilians at 683,000. These represented only modest reductions relative to the QDR targets. In relation to FY 1999, manpower fell by only some 4000 of the remaining 26,000 reduction to active forces needed to achieve the QDR target, and by 5000 of the remaining 36,000 needed to achieve the QDR's reserve target. Nevertheless, civilian end strength fell by 21,000 from FY 1999 to FY 2001 out of a total planned reduction of 64,000 relative to FY 1999 levels. In fact, each service—with the exception of the Air Force—was expected to achieve its active-duty manpower reductions by 2003; selected reserves would be

---

[36]These strikes were not in fact executed until Operation Desert Fox in December 1998. Phoenix Scorpion II, Air Mobility Command's deployment support, also took place in February 1998.

[37]Indeed, many QDR targets had been reached by the end of FY 1999.

Table 4.3
Planned DoD Personnel End-Strength Levels, FY 1998–2003
(in thousands)

	1998 Estimate	1999 Projection	2003 Projection	QDR Goal
Army	488	480	480	480
Navy	387	373	369	369
USMC	173	172	172	172
Air Force	372	371	344	339
Total active	1420	1396	1365	1360
Selected reserves	886	877	837	835
Total civilians	770	747	672	640

SOURCE: Stephen Daggett, *Defense Budget Summary for FY 1999: Data Summary*, Washington, D.C.: Congressional Research Service Report 98-155F, updated June 10, 1998.

only 2000 short of the QDR targets by that time, while civilian personnel would need to be reduced by another 32,000.

The Air Force aimed to achieve its manpower reductions principally through aggressive competitive outsourcing of certain functions (25,400 personnel, primarily those performed by personnel assigned to infrastructure activities), the restructuring of combat forces (4800 personnel), and the streamlining of headquarters (800 personnel).

As Figure 4.4 shows, the execution of the QDR over FY 1997–1999 resulted in only modest reductions to Air Force force structure, manpower, and infrastructure beyond those that already had been accomplished by the Base Force, the Clinton transition year of FY 1994, and the BUR; these reductions were generally in line with the expectation that the QDR generally was a balancing exercise. Nevertheless, some additional reductions were achieved in active-component airlift and active-component, ANG, and civilian personnel, and one TFW was transferred from the active Air Force to the reserve component as planned.

From the Air Force's perspective, the end to further reductions to force structure can be judged a good outcome, as the Air Force did not again have to trade force structure for increasingly elusive modernization. Further, the freedom to choose its own preferred means

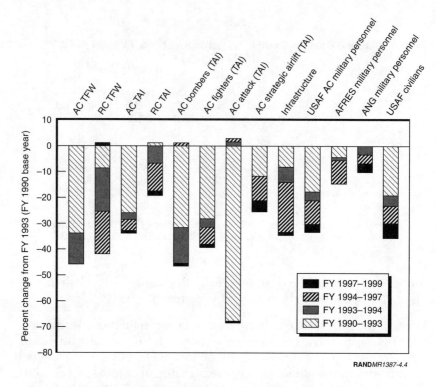

Figure 4.4—USAF Force Structure and Manpower Reductions,
FY 1990–1999 (FY 1990 base year)

of meeting manpower reduction targets (outsourcing) offered some opportunities to undertake reengineering in a creative way even if the result turned out to be somewhat less successful than had been hoped. Nevertheless, as described earlier, the QDR's failure to undertake a more penetrating examination of roles, missions, and force structure, and its desire to use the theater air program as a bill payer, were the source of some frustration.

Defense Reform and Infrastructure

Although the 1995 BRAC process had yielded an estimated $1.8 billion in annual recurring savings, the DoD's hopes to initiate two additional rounds of base closures in 1999 and 2001 were not realized owing to congressional dissatisfaction with the conduct of that round. As a result, by 2000 the DoD had excess infrastructure capacity of some 20 percent, with the Air Force's excess capacity estimated at roughly 25 percent.

Modernization[38]

In terms of recapitalization, most of the programs advocated as part of selective modernization in the QDR have continued at reasonably robust levels, with some programs receiving more funding than initially indicated.

For example, the number of JSTARS was raised from 13 to 16, while the number of F-22 Raptors was raised to 341, three more than indicated in the QDR. A total of 458 MV-22 Ospreys were planned, 98 aircraft more than indicated in the QDR. In line with QDR objectives, a total of 548 F/A-18E/F Super Hornets were planned, while the Joint Strike Fighter program was funded at $23.2 billion. Funds were also directed toward such programs as Deep Strike/Anti-Armor Weapons and Munitions, Nimitz-class aircraft carriers, theater ballistic missile defense, and national missile defense.

Nonetheless, sustaining the necessary level of procurement of defense weapon systems remained a challenge in the tight budgetary environment. According to the General Accounting Office, long-term modernization plans remain at risk, and post-QDR procurement spending plans have been reduced as funding continues to migrate from modernization to operations accounts—precisely the problem that the QDR aimed to resolve.[39] Not only has the DoD continued to place a higher priority on current obligations than on future ones, but procurement spending remains approximately $27 billion below projected needs.[40]

---

[38]Our assessment of the execution of the modernization decisions taken in the QDR is based largely on an examination of acquisition plans as of June 30, 2000.

[39]See U.S. General Accounting Office, *Future Years Defense Program: Risks in Operation and Maintenance and Procurement Programs*, Washington, D.C.: GAO-01-33, October 2000, especially pp. 17–21.

[40]See Congressional Budget Office, *An Analysis of the President's Budgetary Proposals for Fiscal Year 2000*, Washington, D.C., April 1999.

Even more problematic have been the prospects for the sort of force transformation considered under what is conventionally called the "Revolution in Military Affairs" (RMA). One of the QDR's principal decisions regarding modernization was that the transformation of the U.S. military should in fact take place at a somewhat evolutionary pace.[41] It seems clear that the slow rate of transformation that the QDR chose was in large part a function of the relatively modest resources that were available for such activities. Using funding in support of Joint Vision 2010 defense technology objectives as a crude measure of the level of defense resources allocated to transformation, the FY 1999 program suggested that only some $766 million of DoD budget authority was devoted to developing future warfighting capabilities. As a share of defense aggregates, this represented only 2.1 percent of the $36 billion in DoD RDT&E spending requested for FY 1999—or nine-tenths of one percent (0.9 percent) of the approximately $85 billion in investment (procurement and RDT&E) spending and less than three-tenths of one percent (0.3 percent) of the $257 billion in total budget authority. This is considerably less than the $5 billion to $10 billion for transformation that was advocated by the National Defense Panel (NDP) on top of the QDR's $60 billion procurement goal.[42]

Resources

Top Line. The QDR had anticipated that the annual DoD budget would remain at roughly $250 billion a year in constant FY 1997 dollars. With the submission of the FY 1999 budget—the first to implement the results of the QDR—budget authority was in fact pegged at levels just shy of this target (see Table 4.4).

As a result of growing evidence of readiness problems and funding shortfalls, in 1999 the administration proposed increasing long-term

---

[41] This seems reflective of a conception of the RMA that views a true RMA as unlikely but accepts that there will be continuing evolution in equipment, organizations, and tactics to adjust to changes in technology. See Theodor W. Galdi, *Revolution in Military Affairs? Competing Concepts, Organizational Responses, Outstanding Issues*, Washington, D.C., Congressional Research Service Report 95-1170F, December 11, 1995.

[42] The $5 billion to $10 billion estimate is from CJCS Henry Shelton's testimony before the Senate Armed Services Committee, February 3, 1998.

Table 4.4
FY 1999 Long-Range Forecast for DoD Spending

	1997	1998	1999	2000	2001	2002	2003
Then-year $B							
BA	258.0	254.9	257.3	262.9	271.1	274.3	284.0
Outlays	258.3	251.4	252.6	255.8	257.1	259.7	275.8
FY 1997 $B							
BA	257.9	249.3	246.6	246.5	248.7	246.0	248.9
Outlays	258.3	246.0	242.4	240.3	236.2	233.4	242.7

SOURCE: DoD Comptroller, National Defense Budget Estimates, FY 1999, Washington, D.C., March 1998.

defense spending plans. This was for the most part accomplished through the submission of the FY 2000 defense budget and program, which added about $110 billion to long-term defense funding.[43] Figure 4.5 portrays graphically the increases to budget authority that occurred with both the FY 2000 and the FY 2001 budgets.[44]

Like the BUR, the QDR had underestimated the resources that would ultimately be needed—and would ultimately be made available—for defense. Rather than remaining at a level of roughly $250 billion a year in constant FY 1997 dollars as the QDR assumed (see Table 4.5), actual DoD budget authority has exceeded $260 billion a year in constant 1997 dollars since FY 1999.

As a result of these increases to the DoD top line, FY 1998 saw the end of real reductions to defense budget authority over the decade of

---

[43] The $110 billion increase was proposed with the FY 2000 budget with the proviso that Social Security reform was enacted; by the FY 2001 budget, the increase in spending was estimated at $112 billion. As described by OMB, of this $110 billion, $83 billion was for the five years of FY 2000–2004, which included a $64 billion increase above previously planned spending levels and $19 billion in savings from lower inflation and other budgetary and technical adjustments; the remaining $27 billion was said to reflect an increase over the six-year FYDP. See Office of Management and Budget, *Budget of the United States Government, Fiscal Year 2000*, Washington, D.C., 1999, p. 153.

[44] As a point of comparison, between the FY 1997 and FY 1999 President's Budget requests, approximately $18 billion was added for the six years of FY 1997–2002, and between the FY 1999 and FY 2001 requests an additional $80 billion was requested for the five years of FY 1999–2003.

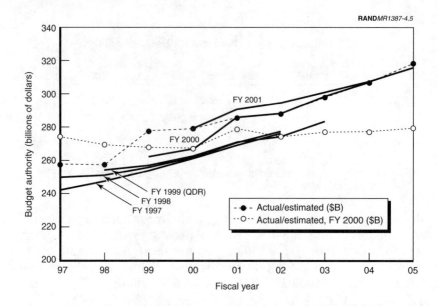

Figure 4.5—DoD Plans for Budget Authority, FY 1997–2001 Budget Requests

Table 4.5

DoD Budget Plans, FY 1998–2001 (BA in billions of dollars)

DoD Plan	1997	1998	1999	2000	2001	2002	2003	2004	2005
FY 1998 (2/97)	250	251	256	263	270	277			
FY 1999 (2/98)		255	257	263	271	274	284		
FY 2000 (2/99)			263	267	286	288	299	308	
FY 2001 (2/00)					280	291	295	301	308	316
Actual BA:									
Then-year $B	258	259	278	280					
FY 1997 $B	258	253	267	263					

SOURCES: Office of Management and Budget, *Budget of the United States Government: Historical Tables*, fiscal years 1998 through 2001, and Department of Defense, *Annual Report to the President and Congress*, Washington, D.C., January 2001, Table 17-1, "Department of Defense Budget Authority," p. 244.

the 1990s. In that year, the cumulative decline in defense budget authority since FY 1990 reached 27.5 percent and then recovered somewhat over the next several years, apparently stabilizing in the 24 to 25 percent range—roughly the real reductions that had originally been anticipated in connection with the Base Force. Real declines in defense spending for the services generally bottomed out at about 33 percent and recovered slightly to reach the 29 to 31 percent range.[45]

Modernization. Procurement spending in FY 1999 was to increase in real terms both as a share of total DoD budget authority and on a per-troop basis. While the FY 1999 budget generally met or exceeded QDR goals for procurement spending and placed the defense program on track to hit the QDR's $60 billion modernization goal (Table 4.6), planned spending on procurement in the FY 1999 program and budget was actually less than that planned in the FY 1998 DoD budget. Meanwhile, the long-range spending plans for RDT&E were revised upward after FY 1998, although each anticipated real RDT&E spending declines in the out years (Table 4.7). The result of these changes is that the share of DoD budget authority devoted to investment was expected to increase modestly from about 33 to 37 percent, while O&S spending was to fall from roughly 66 to 63

Table 4.6

DoD Procurement Plans, FY 1998–2001 (BA in billions of dollars)

DoD Plan	1997	1998	1999	2000	2001	2002	2003	2004	2005
QDR goal			49	54	60	61	62		
FY 1998 (2/97)	44	43	51	57	61	68			
FY 1999 (2/98)		45	49	54	61	61	64		
FY 2000 (2/99)			49	53	62	62	67	69	
FY 2001 (2/00)				54	60	63	67	68	71
Actual BA:									
Then-year $B	43	45	49	53					
FY 2000 $B	45	46	50	53					

SOURCES: Office of Management and Budget, *Budget of the United States Government: Historical Tables*, fiscal years 1998 through 2001, and Office of the Assistant Secretary of Defense (Public Affairs), "Department of Defense Budget for FY 1999," News Release No. 026-98, February 2, 1998.

---

[45]The explanation for the smaller decline in DoD budget authority is found in the substantial real increases to defense-wide spending over the period.

Table 4.7

DoD RDT&E Plans, FY 1998–2001 (BA in billions of dollars)

DoD Plan	1997	1998	1999	2000	2001	2002	2003	2004	2005
FY 1998 (2/97)	37	36	35	33	33	34			
FY 1999 (2/98)		37	36	34	33	34	34		
FY 2000 (2/99)			37	34	34	35	35	35	
FY 2001 (2/00)				38	38	38	38	37	36
Actual BA:									
Then-year $B	36	37	37	34					
FY 2000 $B	38	38	37	34					

SOURCE: Office of Management and Budget, *Budget of the United States Government: Historical Tables*, fiscal years 1998 through 2001.

percent.[46] Investment was expected to stabilize at approximately 30 percent below FY 1990 levels and O&S to stabilize at some 21 percent below 1990 levels.[47]

The View from the Air Force. The QDR called for modest real increases in Air Force spending over FY 1999–2003; as Table 4.8 shows,

Table 4.8

Air Force Spending Plans, FY 1998–2001 (BA in billions of dollars)

DoD Plan	1997	1998	1999	2000	2001	2002	2003	2004	2005
FY 1998 (2/97)	74	75	75						
FY 1999 (2/98)		74	77	78	81	83	85		
FY 2000 (2/99)			77	79	85	87	89	92	95
FY 2001 (2/00)				81	85	88	89	91	93
Actual BA:									
Then-year $B	73	76	82	81					
FY 2000 $B	82	80	82	84					

SOURCE: Office of Management and Budget, *Budget of the United States Government: Historical Tables*, fiscal years 1998 through 2001.

---

[46]See DoD Comptroller, *National Defense Budget Estimates, FY 2000/2001*, Washington, D.C, 1999.

[47]Ibid.

The 1997 Quadrennial Defense Review: Seeking to Restore Balance 111

these increases have been realized,[48] with the result that Air Force budget authority is expected to plateau at roughly $85 billion to $87 billion in constant FY 2000 dollars over FY 2001–2005.

Turning to modernization, Figure 4.6, which describes planned and actual procurement spending by the Air Force, shows that actual spending on procurement generally fell below planned levels up to the FY 1998 plan but has since been generally consistent with plans for FY 1999 and 2000.

Air Force spending on aircraft procurement shows a similar if somewhat less dramatic pattern, as shown in Figure 4.7. Finally, planned and actual Air Force RDT&E spending can be seen to have been substantially below planned levels until about FY 1997 but has

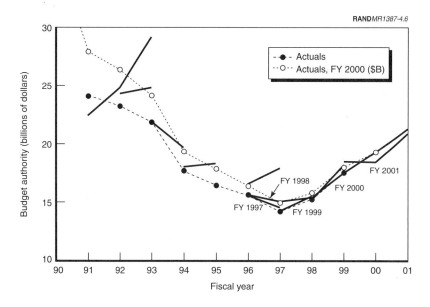

Figure 4.6—Planned and Actual BA, Air Force Procurement

---

[48]See Robert S. Dudney, "Air Force Programs at the Core," *Air Force Magazine*, Vol. 80, No. 6, June 1997. Actual/estimated spending is from DoD Comptroller, *National Defense Budget Estimates, FY 2001*, Washington, D.C., 2000.

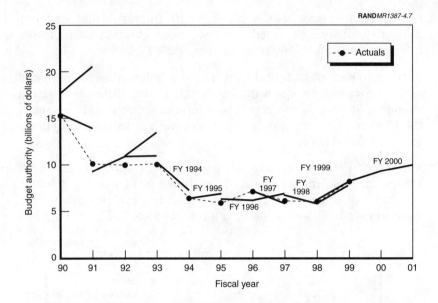

Figure 4.7—Planned and Actual BA for Aircraft Procurement, Air Force

generally tracked with plans since (Figure 4.8), and a decline in USAF RDT&E spending is anticipated over coming years.[49] Thus, over the FY 1997–1999 period, Air Force investment spending recovered somewhat from its cumulative decline since FY 1990, and investment's share of total Air Force budget authority increased slightly, with procurement growing the most.[50]

ASSESSMENT

Capability to Execute the Strategy

The QDR reaffirmed that U.S. forces needed to be able to execute two nearly simultaneous MTWs with moderate risk.[51] As described in

---

[49] The planned future decline in RDT&E spending may in part be related to the maturation of programs such as the F-22, which will be moving from engineering and manufacturing development (under RDT&E) to production (under procurement).

[50] See DoD Comptroller, *National Defense Budget Estimates, FY 2000/2001*, 1999.

[51] Department of Defense, *Report of the Quadrennial Defense Review*, p. 23.

The 1997 Quadrennial Defense Review: Seeking to Restore Balance 113

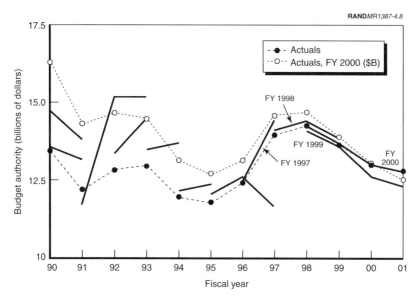

Figure 4.8—Planned and Actual BA for RDT&E, Air Force

the DoD's *Quarterly Readiness Reports*, however, while U.S. forces were capable of executing this strategy, the level of risk associated with it has risen since 1998 (the last year programmed and budgeted under the BUR) and has since been judged on numerous occasions to be in the "moderate-to-high" or "high" range.

In February 1998, CJCS Shelton reported: "While we are undeniably busier and more fully committed than in the past, the U.S. military remains fully capable of executing the National Military Strategy with an acceptable level of risk."[52] By February 1999, however, Chairman Shelton was reporting that the risk associated with execution of the two-conflict strategy had increased in the last quarter of FY 1998 and in the first quarter of FY 1999:

> As I told the Senate Armed Services Committee last September [1998] and again in January [1999], we remain fully capable of

_____
[52]See testimony of Chairman Shelton before the Senate Armed Services Committee, February 3, 1998.

executing our current strategy. As I highlighted in those hearings, however, the risks associated with the most demanding scenarios have increased. We now assess the risk factors for fighting and winning the 1st Major Theater War (MTW) as moderate and for the 2d MTW as high.

As I have explained in the past, this does not mean that we doubt our ability to prevail in either contingency. We are not the "hollow" force of the 1970s, a force that I served in and know well. Nevertheless, increased risk translates into longer timelines and correspondingly higher casualties, and thus leads to our increasing concern.[53]

The change in execution risk for the second conflict was related in part to continued risks associated with strategic mobility shortfalls and in part to eroding readiness, discussed next.

Readiness

Evidence of readiness problems began to accumulate after March 1996. By February 1998, Chairman Shelton, in testimony before the Senate Armed Services Committee, was describing an emerging picture of readiness problems driven by high operational tempo:

There is no question that more frequent deployments affect readiness. We are beginning to see anecdotal evidence of readiness issues in some units, particularly at the tactical level of operations. At the operational and strategic levels, however, we remain capable of conducting operations across the spectrum of conflict.[54]

Although the QDR seems to have rejected a policy of "tiered" readiness,[55] by May 1998 the DoD had begun to report what amounted to such a tiering in readiness: The readiness of U.S. forward and first-to-fight forces was judged as high, with these forces ready and pre-

---

[53]See statement of CJCS Shelton before the House Armed Services Committee, February 2, 1999.

[54]See testimony of Chairman Shelton before the Senate Armed Services Committee, February 3, 1998.

[55]See Department of Defense, *Report of the Quadrennial Defense Review*, p. 36.

pared to execute the national military strategy, including two nearly simultaneous MTWs. At the same time, however, problems had emerged with forces recovering from deployments and training up for their next deployment. Increasing anecdotal evidence of problems with forces further down the deployment chain was given by the Army, while the Navy reported readiness gaps between forward-deployed forces and those that were not forward-deployed and the Marine Corps noted that equipment readiness was slipping.

For the Air Force, high operational tempos were resulting in both short- and long-term readiness problems:

> As we go into '99, our concerns that continue with us in the Air Force are the tempo—we're at a very high tempo. The Air Force transition[ed] from a Cold War force of fairly good size, equivalent to about 36 fighter wings. We've reduced our force structure and completed that by about a third. We reduced our overseas force structure by about two-thirds. At the same time our contingency tasking operations have increased by a factor of four. That drives tempo.
>
> [T]he aging aircraft that I mentioned. We're concerned about that as it continues on because of [the] need to replace not only parts, but also engines and other expensive items to keep that fleet going as we move into our modernization period.
>
> We're right now forecasting about an 1,800 pilot shortfall by '02. That's from a baseline of about 14,200 on our requirement. . . . I would like to be able to say [that it's as bad as it's going to get on retention of pilots and other personnel], but I don't think we're going to get better.[56]

By September 1998 (the last month of the last year guided by the BUR) the readiness problems had become serious and systemic, leading to a meeting with the president in advance of the Joint

---

[56]See Office of the Assistant Secretary of Defense (Public Affairs), background briefing on military readiness, May 5, 1998, statement attributed to a "senior military official" from the Air Force.

Chiefs' readiness testimony to Congress.[57] On September 29, 1998, Chairman Shelton testified that:

> Right now the force is fundamentally sound, but the warning signals cannot and should not be ignored. Let me use an aviation analogy to describe our current situation. In my view, we have "nosed over" and our readiness is descending. I believe that with the support of the Administration and Congress, we should apply corrective action now. We must "pull back on the stick" and begin to climb before we find ourselves in a nosedive that might cause irreparable damage to this great force we have created, a nosedive that will take years to pull out of.[58]

Chairman Shelton concluded, in effect, that the balance the QDR had struck between strategy, forces, and resources had not successfully resolved the major challenges facing the DoD, requiring both constant attention from senior leaders and hard choices in the future.

With modernization and readiness problems emerging, Secretary Cohen testified before the Senate Armed Services Committee on October 6 that the president had directed him to "fix" those areas in the FY 2000 budget:

> It's important we send the signal to the men and women in uniform that we care about them, that we have indeed identified the nature of the problems, and now we've got to take constructive actions to deal with them.[59]

Secretary Cohen reported that he had made a political judgment in establishing the QDR's and the FY 1999 budget's assumptions of a flat, $250 billion-a-year budget and that legislators would not increase defense resources in light of the balanced budget agreement between Congress and the executive branch.[60] Secretary Cohen also

---

[57] See White House, "Remarks by the President to the Joint Chiefs of Staff and Commanders in Chief of the U.S. Armed Forces," September 15, 1998.

[58] See testimony of Chairman Shelton on the readiness of U.S. Armed Forces before the Senate Armed Services Committee, September 29, 1998.

[59] See Jim Garamone, "Fixing the Fiscal 2000 Defense Budget," American Forces Information Service, October 8, 1998.

[60] Ibid. See also "Summing Up: Cohen," *The NewsHour with Jim Lehrer*.

reported that while the FY 1999 budget had more funding for procurement, the department could not reach its $60 billion spending target unless it was allowed to close more bases or overspend.[61]

As a result, the president in November 1998 released $1.1 billion in military readiness funding made available through the Omnibus Consolidated Emergency Supplemental Appropriations Act of 1999, and in December he announced a 4.4 percent military pay increase for FY 2000. In January, the president announced additional measures to address readiness and other problems, the centerpiece for which was a $112 billion increase in defense resources.[62]

Modernization

The centerpiece of the QDR was rebalancing the defense budget and program to free resources to prepare for future threats—and, as described above, the planning and execution of the defense program suggests that the QDR's nominal modernization goal of $60 billion a year in procurement spending is being met. Nevertheless, procurement plans have continued to be trimmed as a result of ongoing migration of funding to operations accounts, and long-term modernization plans remain at risk.[63]

Further, and perhaps more significantly, there are good reasons to believe that the modernization goal itself is problematic. The $60 billion-a-year procurement spending goal for selective modernization of the force, for example, appears to fall well short of what will be required to recapitalize the current force, and by any measure,

---

[61] For additional discussion of the FY 2002–2007 budget and program along with consideration of defense funding in an environment of a federal budget surplus, see the appendix.

[62] See Steven Lee Myers, "Military Leaders Make Case to Clinton for More Money," *New York Times*, September 16, 1998; Steven Lee Myers, "Clinton Is Seeking More Money for Military Readiness," *New York Times*, September 23, 1998; Eric Schmitt, "Joint Chiefs Tell Lawmakers Pet Projects Impair Defense," *New York Times*, September 30, 1998; Steven Lee Myers, "Pentagon's Hopes for Major Budget Increase Wane, but Officials Vow to Fight," *New York Times*, December 14, 1998; and Steven Lee Myers, "Clinton Proposes a Budget Increase for the Military," *New York Times*, January 2, 1999.

[63] See U.S. General Accounting Office, *Future Years Defense Program: Risks in Operation and Maintenance and Procurement Programs*, pp. 17–21.

funding for transformation activities remains quite low. This suggests that the QDR, like the BUR, assumed an exceedingly high discount rate applied to future threats and investments. Taken together, the QDR did not successfully resolve either the DoD's modernization problems or the imbalance in the program. Accordingly, these issues will need to be revisited in the 2001 defense reviews.

SECTION CONCLUSIONS

Although the threat environment had stabilized somewhat and by some measures was even more benign than during the BUR, the QDR continued to be plagued by many of the same problems that had been encountered in the execution of the BUR.[64] Perhaps the most important of these was the QDR's failure to resolve the fundamental mismatch between strategy, forces, and resources.

As with the BUR, the QDR's recommended budget of $250 billion a year clearly underestimated the resources that would actually be required to simultaneously support the strategy and forces without allowing readiness levels to erode further, warfighting costs to increase, or force modernization to be underfunded. Entirely unanticipated by the QDR, additional defense resources were needed and would be made available. Although the implementation of the QDR has not yet been completed,[65] there are some indications that these additional resources have halted—and have possibly begun to reverse—recent unfavorable trends.[66]

---

[64]Indicators of a more benign environment at the time of the QDR include a reduction in the number of major conflicts and refugees; dramatic reductions in worldwide military spending, with an increasing share of spending accounted for by the United States, its allies, and its friends; and a reduction in the inventories of high-performance fighter aircraft, tanks, and other systems.

[65]At the earliest, the execution of the 1997 QDR will probably not be completed until the end of FY 2001. At this time, the new administration's transitional (FY 2002) budget and program will begin to influence the defense program, although early press reports suggest that the administration will make few changes to the existing budget for FY 2002. Given the magnitude of the task of revising the FY 2002 DoD budget in the spring of 2001, implementation of the 2001 QDR will probably not be completed until after FY 2002, at which time the first budget implementing the results of QDR 2001 (FY 2003) will be executed.

[66]See Joint Chiefs of Staff, "Budget Update and Readiness Implications," briefing, March 2001. Nevertheless, O&S costs were growing at a greater rate than expected.

To the extent that they mitigate resource-related shortfalls, the FY 2001 (and 2002) defense programs may indeed reduce the risk level inherent in the two-conflict strategy to the low-to-moderate range[67] and may thus be sufficient to halt or even reverse the apparent erosion in readiness. If this were to happen, the QDR can probably be judged to be at least a partial success. If the increased funding proves insufficient, however, these problems will be left to the next QDR to tackle.

At the same time, the evidence that has accumulated to date suggests that the QDR generally failed to accomplish its main objectives of providing a blueprint for a strategy-based, balanced, and affordable defense program within the assumed annual budget of $250 billion a year and placing the DoD on a path that would lead to modernization of the force.[68]

First, as discussed in this section, the QDR underestimated the resources necessary to fund the overall defense program and, in particular, to support modernization—including both the recapitalization and the transformation of the force. The funding issue was addressed in part by the infusion of an additional $63 billion over FY 1999–2001.[69] However, the costs of recapitalizing the current force are now estimated to be in the $90 billion range, and resources for the transformation of the force remain tiny percentages of budgetary aggregates.[70]

---

[67] It is not at all clear at the time of publication that the new administration intends to continue using the two-MTW construct for force sizing.

[68] As the General Accounting Office recently observed, "[A] mismatch exists between Defense's plans and the projected available funding. Optimistic planning provides an unclear picture of defense priorities because tough decisions and trade-offs between needs and wants are avoided." See U.S. General Accounting Office, *Future Years Defense Program: Risks in Operation and Maintenance and Procurement Programs*, p. 22.

[69] See Department of Defense, *Annual Report to the President and Congress*, Washington, D.C., 2001, Table 17-1, "Department of Defense Budget Authority," p. 244.

[70] See William Cohen, testimony of Secretary Cohen before the House Armed Services Committee, February 9, 2000; Congressional Budget Office, *Budgeting for Defense: Maintaining Today's Forces*, Washington, D.C., September 2000; and Joint Chiefs of Staff, "Budget Update and Readiness Implications."

Second, the QDR's inability to halt the continued migration of resources from modernization to operations accounts remains a problem, with the result that the DoD continues to have difficulties meeting its planned growth in procurement funds. As planned procurement funding has been reduced, modernization plans have been shifted to the future.[71]

Finally, the chronic underfunding and increasing backlogs of depot and real property maintenance, the backlog of needed military construction, the underfunding of the Defense Health Program, and other related problems indicate yet another symptom or consequence of the current imbalance: the postponement of all but the most immediately critical spending.[72]

The next chapter provides some concluding observations.

---

[71] See U.S. General Accounting Office, *Future Years Defense Program: Risks in Operation and Maintenance and Procurement Programs*, pp. 6–7.

[72] Ibid.

Chapter Five
CONCLUSIONS

This history of the three major defense strategy reviews of the last decade aimed to highlight the inputs (assumptions, threats, and domestic environments), outputs (decisions and other outcomes), and implementation experience of each review. After identifying some common features of the reviews, we offer some lessons regarding strategy, forces, and budgets, and we then close with some thoughts on how defense planning might be improved.

Stepping back from their details, the reviews appear to have shared at least three main features, each of which could benefit from additional scrutiny:

- First, each assumed that the most important (and taxing) mission for conventional forces was halting and reversing cross-border aggression by large-scale mechanized forces. The experience in Kosovo suggests that adversaries have adapted to avoid Desert Storm–style outcomes, and the decline in mechanized forces worldwide raises questions about the continued utility of emphasizing this sort of scenario.

- Second, each review in its own way treated presence and smaller-scale peace and other contingency operations as "lesser-included cases" that could successfully be managed by a force structure designed primarily for warfighting, and each assumed that these contingency operations would impose minimal costs and risks for warfighting. Recent experience suggests that these assumptions are true only when such operations are incidental and short-lived, as was the general pattern during much of the Cold War; by contrast, when such operations are large, tend to accumulate, and need to be sustained over time, they can in fact be quite taxing on warfighting capabilities, affecting readiness and increasing strategic risk.

- Third, each review suffered from the absence of a bipartisan consensus on a post–Cold War foreign and defense policy, and this made the gaps that emerged between strategy, forces, and budgets particularly salient while arguably impeding their successful resolution. The new administration should consider how best to establish a shared vision of the nation's defense priorities, a better partnership with Congress, and a process for fuller consideration of defense funding needs.

We now turn to some lessons for strategy, forces, and budgets.

Regarding *strategy*, the historical record suggests that it is critically important to understand that changes in strategy—a regular feature of presidential transitions and defense reviews—can have a range of important ramifications. The change in normative aims and conception of engagement pursued by the Clinton administration and documented in the BUR, for example, underscored the importance of ethnic conflict and civil strife, promoted peace operations as a more important tool of U.S. policy, and had strong implications for the resulting pattern of U.S. force employment. Having failed to fit force structure and budgets to strategy, the resulting effects could and should have been better anticipated and resources realigned to mitigate or eliminate them.

Another critical result has to do with *force structure*. Table 5.1 shows that while there have been substantial reductions in force structure and manpower, only a modest amount of reshaping of the force has actually taken place. Efforts to meaningfully modernize and transform the force have been hampered by a high discount rate that has elevated current-day threats, force structure, and readiness concerns while effectively discounting longer-term needs. As witnessed by the remarks of former Air Force Chiefs of Staff McPeak and Fogleman, this has been a source of great disappointment for Air Force leaders, who hoped that these reviews might include a more penetrating examination of roles, missions, and force restructuring.

Table 5.1

Proposed Force Structure Changes: Base Force, BUR, and QDR

	FY 1990	1997 Base Force	1999 BUR Force	2003 QDR Force	FY 2001
Air Force					
TFWs (AC/RC)	24/12	15.3/11.3	13/7	12+/8	12+/7+
Bombers (active)	228	181	184	187	181
Land-based ICBMs	1000	550	550	550	550
Navy					
Aircraft carriers	15/1	12/1	11/1	11/1	12/0
Battle force ships	546	448	346	306	317
Marines					
Divisions (AC/RC)	3/1	3/1	3/1	3/1	3/1
Army					
Divisions (AC/RC)	18/10	12/8[a]	10/5+	10/8	10/8
End strength					
Active duty	2070	1626	1418	1360	1382
Reserve	1128	920	893	835	864

[a]RC includes two cadre divisions.

With regard to *budgets*, there seems to have been a chronic reluctance to acknowledge what reasonable-risk versions of a strategy and force structure might really cost. While gaps between strategy, force structure, and resources are not unprecedented,[1] the tacit agreement of the executive and legislative branches to avoid debates over a higher defense top-line, as well as fundamental issues of strategy and policy, may actually have impeded full disclosure and consideration of the problems that plagued the defense program for much of the decade. Instead, the reliance on modest year-to-year revisions that did not upset discretionary spending limits, coupled with the recurring exploitation of the loophole provided by emergency supplementals to mitigate particularly acute shortfalls, meant that the debates would occur only at the margin, and at a rhetorical level only. Failure to tackle these issues head on may have retarded the recog-

---

[1]During the Cold War period, for example, airlift capacity remained well short of the 66 MTM/D that was the stated requirement for responding to a Soviet/Warsaw Pact attack across the inter-German border and a Soviet invasion of Iran. Current military airlift capacity is judged to be nearly 20 percent short of the requirement established by the 1995 MRS BURU and roughly 33 percent short of the requirement established in the more recent MRS05.

nition and remediation of the growing gaps between strategy, forces, and resources.

The Air Force arguably did quite well in recognizing what it could contribute in the post–Cold War environment and was trained and equipped to support the operations it was called on to undertake. Nevertheless, it did less well recognizing signposts for opportunities and challenges, recognizing events whose outcomes it did not control, and positioning itself best in those circumstances. It failed in its efforts to press for a more searching reexamination of roles, missions, and options for restructuring U.S. forces for the post–Cold War era.

Shifting to the present, the new administration's defense review will wrestle with the same questions its predecessors faced: What are to be the nation's aims in the world? What are the main threats and opportunities it faces? What strategy and force structure will best serve the interests of the nation? What resources are needed to ensure low-to-moderate execution risk in that strategy, and capable and ready forces, both now and over the next 20 to 30 years?

In answering these questions, the new administration will be no more encumbered by the assumptions and decisions of the 1997 QDR than the incoming Clinton administration was by those of the Base Force. Nevertheless, the Department of Defense—and the Air Force—would profit from an assumption-based planning approach in which signposts are established that can be used to gauge whether the key assumptions on which planning is predicated are still justified.[2] These include assumptions about future threats, the likely frequency and mix of future missions, the adequacy of forces to undertake these missions, effects on overall readiness and strategic risk, and the availability of resources.

Such an approach is important by virtue of another great lesson of the past decade: that failure to recognize and respond promptly and effectively to emerging gaps and shortfalls can lead to the greatest

---

[2] See James A. Dewar, Carl H. Builder, William M. Hix, and Morlie H. Levin, *Assumption-Based Planning: A Planning Tool for Very Uncertain Times*, Santa Monica: RAND, MR-114-A, 1993.

and most protracted imbalances between strategy, forces, and resources.

Appendix
POSTSCRIPT: FROM DEFICIT POLITICS TO THE POLITICS OF SURPLUS

As this report describes, the choices available to policymakers since 1989 have been both shaped and constrained by the priority accorded to deficit reduction, which effectively placed discretionary defense spending within agreed-upon caps. This appendix provides some final thoughts on the potential the current environment of budgetary surplus may afford.

FROM DEFICIT TO SURPLUS, FY 1981–1998

As Table A.1 shows, each of the fiscal years 1981–1997 saw a federal budget deficit, although the annual deficit declined steadily from its peak in FY 1992. The principal efforts to reduce the deficit by controlling discretionary spending—comprising roughly one-third of total federal spending—found expression in a number of public laws and bills over the last decade (see Table A.2). As shown in Table A.3, most of the "balancing" of the budget was accomplished through reductions to the defense budget; the share of total discretionary spending accounted for by defense discretionary spending fell from 60 percent in FY 1990 to 47 percent in FY 2001.

Nevertheless, although there was general agreement as early as the beginning of 1996 that the defense budget would have to rise, it was not until the fall of 1998—when budget surpluses were first projected and the service chiefs voiced their concerns about readiness problems—that real increases in defense resources were proposed and

Table A.1

Annual Deficit or Surplus, FY 1981–2000 (in billions of dollars)

Fiscal Year	Deficit/Surplus ($B)	As Percentage of GDP	Standardized Budget Deficit/Surplus ($B)	As percentage of GDP
1981	−79	−2.6	−17	−0.5
1982	−128	−4.0	−52	−1.5
1983	−208	−6.0	−120	−3.3
1984	−185	−4.8	−144	−3.7
1985	−212	−5.1	−177	−4.2
1986	−221	−5.0	−212	−4.8
1987	−150	−3.2	−155	−3.3
1988	−155	−3.1	−127	−2.5
1989	−152	−2.8	−115	−2.1
1990	−221	−3.9	−119	−2.1
1991	−269	−4.5	−151	−2.5
1992	−290	−4.7	−184	−2.9
1993	−255	−3.9	−181	−2.7
1994	−203	−2.9	−138	−2.0
1995	−164	−2.2	−136	−1.8
1996	−108	−1.4	−89	−1.1
1997	−22	−0.3	−56	−0.7
1998	69	0.8	−18	−0.2
1999	124	1.4	20	0.2
2000	236	2.4	106	1.1

SOURCE: Congressional Budget Office, *The Budget and Economic Outlook: Fiscal Years 2002–2011*, Washington, D.C., January 2001, p. 139.

realized through the FY 2000 and 2001 President's Budgets.[1] Discretionary spending caps were also relaxed;[2] in its FY 2002 budget proposal, the Bush administration has proposed a further relaxation of discretionary spending caps. The political economy had changed, and the nation had entered the politics of surplus.

---

[1] It is worth noting that since 1998, the growth of total discretionary spending has outpaced that of inflation. See Congressional Budget Office, *The Budget and Economic Outlook: Fiscal Years 2002–2011*, Washington, D.C., January 2001, p. 6.

[2] For example, to accommodate additional discretionary spending in 2001, Congress and the president increased the caps on budget authority and outlays by $99 billion and $59 billion, respectively. Op. cit., p. 75.

Table A.2

Caps on BA and Outlays, FY 1991–2002 (in billions of dollars)

Fiscal Year	Defense Discretionary BA	Defense Discretionary Outlays	Total Discretionary BA	Total Discretionary Outlays
Budget Enforcement Act of 1990 (P.L. 101-508)				
1991	288.9	297.7		
1992	291.6	295.7		
1993	291.8	292.7		
1994			510.8	534.8
1995			517.7	540.8
Omnibus Budget Reconciliation Act of 1993 (P.L. 103-66)				
1994			509.9	537.3
1995			517.4	538.95
1996			519.1	547.3
1997			528.1	547.3
1998			530.6	547.9
FY 1995 Budget Resolution (H. Con. Res. 218)				
1996			−4.0	−5.4
1997			−10.7	−2.4
1998			−4.1	−0.5
FY 1996 Budget Resolution (H. Con. Res. 67-H Rept 104-159)				
1996	265.4	264.0	465.1	531.8
1997	268.0	265.7	482.4	520.3
1998	269.7	264.5	490.7	512.6
1999			482.2	510.5
2000			489.4	514.2
2001			496.6	516.4
2002			498.8	515.1
Balanced Budget Act of 1997 (P.L. 105-33)				
1998	269.0	266.8		(538.8)
1999	271.5	266.5		(538.0)
2000			532.7	558.7
2001			542.0	564.4
2002			551.1	560.8

NOTE: Numbers in parentheses = total of defense and nondefense discretionary spending; does not include violent crime reduction.

SOURCES: "Budget Reconciliation Act Provisions," *1990 CQ Almanac*, Washington, D.C.: Congressional Quarterly Press, 1991, p. 161; "1993 Budget Reconciliation Act," *1993 CQ Almanac*, Washington, D.C.: Congressional Quarterly Press, 1994, p. 139; H. Con. Res. 218; and "GOP Throws Down Budget Gauntlet," *1995 CQ Almanac*, Washington, D.C.: Congressional Quarterly Press, 1996, pp. 2–22.

Table A.3

Defense and Nondefense Discretionary Outlays, FY 1991–2001 (in billions of dollars)

Fiscal Year	Defense Outlays ($B)	Percentage of Total	Nondefense Outlays ($B)	Percentage of Total	Total Outlays ($B)
1991	320	60	214	40	533
1992	303	57	232	43	535
1993	292	54	249	46	541
1994	282	52	262	48	544
1995	274	50	272	50	546
1996	266	50	269	50	534
1997	272	49	277	51	549
1998	270	49	284	51	555
1999	275	48	300	52	575
2000	295	48	322	52	617
2001	301	47	345	53	646

SOURCE: Congressional Budget Office, *The Budget and Economic Outlook: Fiscal Years 2002–2011*, Washington, D.C., January 2001, Table 4.3, p. 75.

THE "FOUR PERCENT SOLUTION"

By 2000–2001, the growing surplus had so fundamentally altered the politics of the fiscal environment that calls were growing for a "four percent solution"—i.e., to raise defense spending to 4 percent (or more) of gross domestic product.[3] From a historical perspective, this would on first inspection appear to be a relatively modest increase (see Figure A.1). However, given the dramatic growth of the U.S. economy over the last decade together with projections of continued growth, the likely result of spending 4 percent of gross domestic product on defense would be a substantial—and even unprecedented—level of real defense spending in peacetime (see Figure A.2), eclipsing even that of 1945, the last year of the Second World War.

---

[3]See, for example, Daniel Gouré and Jeffrey Ranney, *Averting the Defense Train Wreck in the New Millennium*, Washington, D.C.: Center for Strategic and International Studies, 1999; Center for Security Policy, "The 'Four Percent Solution,'" Washington, D.C., CSP Publication No. 00-D72, August 7, 2000; and Hunter Keeter, "Marine Commandant Calls for Defense Spending Increase," *Defense Daily*, August 16, 2000. For a critique, see Franklin C. Spinney, "Madness of Versailles: The 4% Solution," August 20, 2000, http://www.d-n-i.net/FCS_Folder/comments/c381.htm.

Postscript: From Deficit Politics to the Politics of Surplus 131

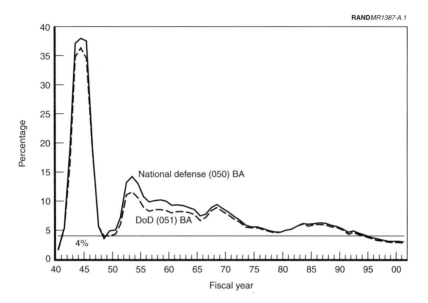

Figure A.1—Defense Aggregates as a Percentage of GDP, 1940–2001

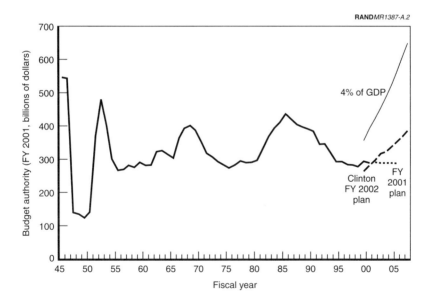

Figure A.2—Notional Budget Consequences of a "Four Percent Solution"

A "four percent solution" could in fact result in an increase of more than $1 trillion in constant FY 2001 dollars over the six years of the outgoing Clinton administration's FY 2002–2007 FYDP, or more than $150 billion a year (see Table A.4). This increase, if continued for another four years, would begin to approximate the ten-year plan for $1.35 trillion in tax cuts that was passed by Congress. Indeed, 4 percent of GDP would consume more than the projected on-budget surplus and most of the total (on- and off-budget) projected budget surplus.[4]

Table A.4

FY 2002 President's Budget Request and "Four Percent Solution"
(discretionary budget authority in billions of dollars)

	Estimated 2001	Projected 2002	2003	2004	2005	2006	2007	2002–2007 Total
Projected GDP ($T)	10.5	11.1	11.7	12.3	13.0	13.7	14.4	
FY 2002 President's Budget Request	296	310	310	317	324	333	342	1936
"Four percent solution"	421	444	468	493	519	547	576	3047
Difference[a]	+125	+134	+158	+176	+195	+214	+234	+1111
In FY 2001 $B[b]								
FY 2002 President's Budget	296	304	297	298	298	300	302	1799
"Four percent solution"	421	435	449	463	478	493	508	2826
Difference[a]	+125	+131	+151	+165	+180	+193	+206	+1027

SOURCES: Department of Defense, Annual Report to the President and Congress, Washington, D.C., January 2001, p. 244, and Office of Management and Budget, FY 2002 Economic Outlook, Highlights from FY 1994 to FY 2001, FY 2002 Baseline Projections, Washington, D.C., January 2001, Table II-1, "Economic Assumptions," p. 24.

[a]Difference between the FY 2002 President's Budget request and the "four percent solution."

[b]Constant FY 2001 billions of dollars, computed by using updated GDP deflators in Office of Management and Budget, FY 2002 Economic Outlook, Highlights from FY 1994 to FY 2001, FY 2002 Baseline Projections, Washington, D.C., January 2001, Table II-1, "Economic Assumptions," p. 24.

---

[4]The Congressional Budget Office projects that in the absence of new legislation, total budget surpluses will grow from some 3 percent to more than 5 percent of GDP from 2002 through 2011. Of this, on-budget surpluses would range from 1 to 3 percent over the 2002–2011 period. See Congressional Budget Office, *The Budget and Economic Outlook: Fiscal Years 2002–2011*, p. xiv.

The initial budget for FY 2002 released by the Bush administration in late February 2001 is something of a placeholder, and the top line for DoD does not differ significantly from the outgoing budget plan of its predecessor for FY 2002 (see Table A.5).[5] Nevertheless, the proposed FY 2002 budget is subject to further revision as the results of the administration's "top-to-bottom" review of the nation's defense needs and QDR 2001 become known—although the full impact of the reviews is unlikely to be felt until the FY 2003 output is submitted in early 2002.

As of late March 2001, little more was known about the details of the defense budget beyond the following: The budget adds $1.4 billion for a military pay raise and allowances, increases by $400 million funding to improve the quality of housing or reduce out-of-pocket housing expenses, and funds new and expanded health benefits for military retirees. It has also been announced that the administration will propose a $2.6 billion research and development initiative for

Table A.5

Comparison of Bush and Clinton FY 2002 Defense Budgets
(discretionary budget authority in billions of dollars)

	2000	2001	2002
National Defense Function			
Discretionary budget authority			
Clinton (Baseline Discretionary BA)	301	311	321
Bush	301	311	325
Department of Defense			
Discretionary budget authority			
Clinton	287	296	310
Bush	287	296	311

SOURCES: Office of Management and Budget, *FY 2002 Economic Outlook, Highlights from FY 1994 to FY 2001, FY 2002 Baseline Projections,* Washington, D.C., January 2001, p. 228; Department of Defense, *Annual Report to the President and Congress,* Washington, D.C., January 2001, p. 244; and Office of Management and Budget, *A Blueprint for New Beginnings: A Responsible Budget for America's Priorities,* Washington, D.C., February 28, 2001, p. 197.

---

[5]Estimates for the out years of FY 2003–2006 simply keep up with inflation.

missile defense alternatives and new technologies to support the transformation of U.S. military capabilities.[6]

Although it is too early to determine whether and how much the new administration will further increase defense budgets in the FY 2002 and FY 2003 defense programs, it seems clear that even in a time of seeming plenitude, the tax cuts will only heighten the competition between defense and other claimants for federal resources.

---

[6]See Office of Management and Budget, *A Blueprint for New Beginnings: A Responsible Budget for America's Priorities*, Washington, D.C., February 28, 2001.

BIBLIOGRAPHY

Adams, Gordon, "Contingencies Serve Role," *Defense News*, April 13, 1998, p. 21.

Armed Forces Information Service, *Defense 97: Commitment to Readiness*, Washington, D.C., 1997.

Aspin, Les, *The Bottom-Up Review: Forces for a New Era*, Washington, D.C., September 1993.

———, *Force Structure Excerpts: Bottom-Up Review*, Washington, D.C., September 1993.

———, *Report on the Bottom-Up Review*, Washington, D.C., October 1993.

———, Secretary of Defense Les Aspin's remarks at the National Defense University graduation, Fort McNair, Washington, D.C., June 16, 1993.

Aspin, Les, and Colin Powell, "Bottom-Up Review," briefing slides, Washington, D.C., September 1, 1993.

Assistant Secretary of the Air Force (Financial Management and Comptroller), *United States Air Force Statistical Digest*, Washington, D.C., various years.

Booz-Allen and Hamilton, *An Assessment of the National Defense Review Process*, McLean, VA, December 1999.

Boutros-Ghali, Boutros, *An Agenda for Peace: Preventive Diplomacy, Peacemaking and Peacekeeping*, New York: United Nations, June 17, 1992.

"Budget Adopted After Long Battle; Five-Year Plan Promises $496 Billion in Deficit Reduction," *1990 CQ Almanac*, Washington, D.C.: Congressional Quarterly Press, 1991, pp. 111–166.

"Budget Reconciliation Act Provisions," *1990 CQ Almanac*, Washington, D.C.: Congressional Quarterly Press, 1991.

Burnette, Thomas N., Jr., Deputy Chief of Staff for Operations and Plans, U.S. Army, testimony before the House Armed Services Committee, Subcommittee on Military Readiness, March 18, 1998.

Center for Security Policy, "The 'Four Percent Solution,'" Washington, D.C., CSP Publication No. 00-D72, August 7, 2000.

Cheney, Richard, testimony before the Senate Armed Services Committee, January 31, 1992.

Clinton, Bill, "A New Covenant for American Security," speech given at Georgetown University, December 12, 1991.

———, "Remarks of Governor Bill Clinton," Los Angeles World Affairs Council, August 13, 1992.

Cohen, William S., testimony of Secretary Cohen before the House Armed Services Committee, February 9, 2000.

Collins, John M., *National Military Strategy, the DoD Base Force, and U.S. Unified Command Plan: An Assessment*, Washington, D.C.: Congressional Research Service Report 92-493S, June 11, 1992.

Congressional Budget Office, *An Analysis of the President's Budgetary Proposals for Fiscal Year 1993*, Washington, D.C., March 1992.

———, *An Analysis of the President's Budgetary Proposals for Fiscal Year 2000*, Washington, D.C., April 1999.

———, *The Budget and Economic Outlook: Fiscal Years 2002–2011*, Washington, D.C., January 2001.

———, *Budgeting for Defense: Maintaining Today's Forces*, Washington, D.C., September 2000.

———, *The Economic and Budget Outlook: An Update*, Washington, D.C., September 1997.

———, *The Economic and Budget Outlook: Fiscal Years 1998–2007*, Washington, D.C., January 1997.

———, *The Economic Effects of the Savings and Loan Crisis*, Washington, D.C., January 1992.

———, *Emergency Spending Under the Budget Enforcement Act: An Update*, Washington, D.C., June 8, 1999.

———, *Fiscal Implications of the Administration's Proposed Base Force*, Washington, D.C., December 1991.

———, *Reducing the Deficit: Revenue and Spending Options*, Washington, D.C., March 1997.

———, *Trends in Selected Indicators of Military Readiness, 1980 Through 1993*, Washington, D.C., March 1994.

Correll, John T., "Two at a Time," *Air Force Magazine*, Vol. 76, No. 9, September 1993.

Daggett, Stephen, *A Comparison of Clinton Administration and Bush Administration Long-Term Defense Budget Plans for FY 1994–99*, Washington, D.C.: Congressional Research Service Report 95-20F, December 20, 1994.

———, *Defense Budget Summary for FY 1999: Data Summary*, Washington, D.C.: Congressional Research Service Report 98-155F, updated June 10, 1998.

Defense Science Board, *Report of the Defense Science Board Task Force on Readiness*, June 1994.

———, *Task Force on the Fiscal Years 1994–99 Future Years Defense Program (FYDP)*, reports of May 3, 1993, and June 29, 1993.

Department of Defense, *Annual Report to the President and Congress*, Washington, D.C., various years.

———, *1992 Joint Military Net Assessment*, Washington, D.C., August 1992.

———, Pentagon Operations Directorate, "Bottom-Up Review Briefing by SECDEF and CJCS," Secretary of Defense Message P020023Z SEP 93, September 2, 1993.

———, *Quarterly Readiness Report to the Congress*, various periods, Washington, D.C., 1999–present.

———, *Report of the Quadrennial Defense Review*, Washington, D.C., May 1997.

———, *Report to Congress on U.S. Military Involvement in Major Smaller Scale Contingencies Since the Persian Gulf War*, Washington, D.C., March 1999.

Deutch, John, "Memorandum for Distribution, Subject: Final Report of the Defense Science Board Task Force on Readiness," July 1994.

Dewar, James A., Carl H. Builder, William M. Hix, and Morlie H. Levin, *Assumption-Based Planning: A Planning Tool for Very Uncertain Times*, Santa Monica: RAND, MR-114-A, 1993.

DoD Comptroller, *National Defense Budget Estimates*, Washington, D.C., various years.

Dudney, Robert S., "Air Force Programs at the Core," *Air Force Magazine*, Vol. 80, No. 6, June 1997.

Ellis, James O., Jr., Deputy Chief of Naval Operations (Plans, Policy, and Operations), testimony before the House Armed Services Committee, Subcommittee on Military Readiness, March 18, 1998.

Fogleman, Ronald, posture statement presented in testimony before the House National Security Committee, March 5, 1997.

Galdi, Theodor W., *Revolution in Military Affairs? Competing Concepts, Organizational Responses, Outstanding Issues*, Washington, D.C., Congressional Research Service Report 95-1170F, December 11, 1995.

Gamble, Patrick K., Deputy Chief of Staff, Air and Space Operations, U.S. Air Force, testimony before the House Armed Services Committee, Subcommittee on Military Readiness, March 18, 1998.

Garamone, Jim, "Fixing the Fiscal 2000 Defense Budget," American Forces Information Service, October 8, 1998.

———, "Shelton Warns of Readiness Problems," American Forces Information Service, October 1, 1998.

"GOP Throws Down Budget Gauntlet," *1995 CQ Almanac*, Washington, D.C.: Congressional Quarterly Press, 1996.

Gordon, Michael, "Cuts Force Review of War Strategies," *New York Times*, May 30, 1993, p. 16.

Gouré, Daniel, and Jeffrey Ranney, *Averting the Defense Train Wreck in the New Millennium*, Washington, D.C.: Center for Strategic and International Studies, 1999.

Gunzinger, Mark, "Beyond the Bottom-Up Review," in *Essays on Strategy XIV*, Washington, D.C.: Institute for National Security Studies, National Defense University, 1996.

Jaffe, Lorna S., *The Development of the Base Force, 1989–1992*, Washington, D.C.: Joint History Office, Office of the Chairman of the Joint Chiefs of Staff, July 1993.

Joint Chiefs of Staff, "Budget Update and Readiness Implications," briefing, March 2001.

———, *Joint Military Net Assessment*, Washington, D.C., 1989 through 1993.

———, *Joint Military Net Assessment*, Washington, D.C., 1994 (not releasable to the general public).

———, *1992 National Military Strategy*, Washington, D.C., January 1992.

Keeter, Hunter, "Marine Commandant Calls for Defense Spending Increase," *Defense Daily*, August 16, 2000.

Kohn, Richard H., "The Early Retirement of General Ronald R. Fogleman, Chief of Staff, United States Air Force," *Aerospace Power Journal*, Spring 2001.

Lacroix, Francis W., testimony before the House Armed Services Committee, March 1, 1994.

Lewis, Leslie, C. Robert Roll, and John D. Mayer, *Assessing the Structure and Mix of Future Active and Reserve Forces: Assessment of Policies and Practices for Implementing the Total Force Policy*, Santa Monica: RAND, MR-133-OSD, 1992.

Matthews, William, "Soviet Demise Leaves Pentagon Wondering Who Is the Foe," *Defense News*, February 24, 1992.

McCain, John, "Defense Preparedness," Congressional Record, Senate, September 30, 1998, pp. S11139–S11142.

———, "Status of Operational Readiness of U.S. Military Forces," *Congressional Record*, Senate, September 10, 1998, pp. S10198–S10201.

Myers, Steven Lee, "Clinton Is Seeking More Money for Military Readiness," *New York Times*, September 23, 1998.

———, "Clinton Proposes a Budget Increase for the Military, *New York Times*, January 2, 1999.

———, "Military Leaders Make Case to Clinton for More Money," *New York Times*, September 16, 1998.

———, "Pentagon's Hopes for Major Budget Increase Wane, but Officials Vow to Fight," *New York Times*, December 14, 1998.

Nash, Colleen A., "Snapshots of the New Budgets," *Air Force Magazine*, April 1991.

"1993 Budget Reconciliation Act," *1993 CQ Almanac*, Washington, D.C.: Congressional Quarterly Press, 1994.

Office of the Assistant Secretary of Defense (Public Affairs), background briefing on military readiness, May 5, 1998.

———, "Department of Defense Budget for FY1999," News Release No 026-98, February 2, 1998.

———, "Quadrennial Defense Review, DoD news briefing, Washington, D.C., May 19, 1997.

———, "Readiness Task Force Presents Its Findings," OASD(PA) News Release No. 437-94, July 22, 1994.

Office of Management and Budget, "A Blueprint for New Beginnings: A Responsible Budget for America's Priorities," Washington, D.C., February 28, 2001.

———, *Budget of the United States Government, Fiscal Year 1992*, Washington, D.C., February 4, 1991.

———, *Budget of the United States Government, Fiscal Year 1993*, Washington, D.C., January 1992.

———, *Budget of the United States Government, Fiscal Year 2000*, Washington, D.C., 1999.

———, *Budget of the United States Government: Historical Tables*, fiscal years 1998 through 2001, Washington, D.C., various years.

———, *FY 2002 Economic Outlook, Highlights from FY 1994 to FY 2001, FY 2002 Baseline Projections*, Washington, D.C., January 2001.

———, *A Vision of Change for America*, Washington, D.C., February 17, 1993.

Perry, William, testimony before the Senate Armed Services Committee, June 17, 1993.

Powell, Colin L., "The Base Force: A Total Force," presentation to the Senate Appropriations Committee, Subcommittee on Defense, September 25, 1991.

———, "Building the Base Force: National Security for the 1990s and Beyond," annotated briefing, September 1990.

———, statement of General Colin Powell, Chairman, Joint Chiefs of Staff, in U.S. Senate, Committee on Armed Services, Department of Defense Authorization for Appropriations for Fiscal Years 1992 and 1993, February 21, 1991.

———, testimony before the House Armed Services Committee, February 6, 1992.

———, "U.S. Forces: Challenges Ahead," *Foreign Affairs*, Vol. 71, No. 5, Winter 1992–1993, pp. 32–45.

Rice, Donald B., "Foreword," in U.S. Air Force, *Toward the Future: Global Reach–Global Power: U.S. Air Force White Papers, 1989–1992*, Washington, D.C., January 1993.

Robertson, Charleston T., statement of General Charleston T. Robertson, Jr., USAF, Commander in Chief, U.S. Transportation Command, before the House Armed Services Committee, March 22, 1999.

Rostker, Bernard, et al., *Assessing the Structure and Mix of Future Active and Reserve Forces: Final Report to the Secretary of Defense*, Santa Monica: RAND, MR-140-1-OSD, 1992.

Ryan, Michael C., *Military Readiness, Operations Tempo (OPTEMPO) and Personnel Tempo (PERSTEMPO): Are U.S. Forces Doing Too Much?* Washington, D.C.: Congressional Research Service Report 98-41F, January 14, 1998.

Schmitt, Eric, "Joint Chiefs Tell Lawmakers Pet Projects Impair Defense," *New York Times*, September 30, 1998.

Schrader, John, Leslie Lewis, and Roger Allen Brown, *Quadrennial Defense Review (QDR) Analysis: A Retrospective Look at Joint Staff Participation*, Santa Monica: RAND, DB-236-JS, 1999.

Serafino, Nina, *Peacekeeping: Issues of Military Involvement*, Washington, D.C.: Congressional Research Service Issue Brief, July 2000.

Shalikashvili, John, Chairman Shalikashvili's testimony before the Senate Armed Services Committee hearings on August 4, 1994.

Shelton, Henry, statement of CJCS Shelton before the House Armed Services Committee, February 2, 1999.

———, testimony of Chairman Shelton before the Senate Armed Services Committee, February 3, 1998.

———, testimony of Chairman Shelton on the readiness of U.S. Armed Forces before the Senate Armed Services Committee, September 29, 1998.

Snider, Don M., *Strategy, Forces and Budgets: Dominant Influences in Executive Decision Making, Post–Cold War, 1989–91*, Carlisle Barracks, PA: Strategic Studies Institute, Professional Readings in Military Strategy No. 8, February 1993.

Spence, Floyd D., *Military Readiness 1997: Rhetoric and Reality*, House Committee on National Security, April 9, 1997.

———, "Statement of Honorable Floyd D. Spence, Fiscal Year 1998 SECDEF/CJCS Posture Hearing," February 12, 1996.

Spinney, Franklin C., "Madness of Versailles: The 4% Solution," August 20, 2000, http://www.d-n-i.net/FCS_Folder/comments/c381.htm.

Steele, Martin R., Deputy Chief of Staff for Plans, Policy, and Operations, testimony before the House Armed Services Committee, Subcommittee on Military Readiness, March 18, 1998.

"Summing Up: Cohen," *The NewsHour with Jim Lehrer*, Public Broadcasting System, January 10, 2001.

Thomason, James S., Paul H. Richanbach, Sharon M. Fiore, and Deborah P. Christie, *Quadrennial Review Process: Lessons Learned from the 1997 Review and Options for the Future*, Alexandria, VA: Institute for Defense Analyses, IDA Paper P-3402, August 1998.

Tirpak, John A. "Projections from the QDR," *Air Force Magazine*, Vol. 80, No. 8, August 1997.

Tyler, Patrick E., "Pentagon Drops Goal of Blocking New Superpowers," *New York Times*, May 24, 1992, p. 1.

U.S. Air Force, *Air Force Strategic Plan, Vol. 2: Performance Plan Annex; Performance Measure Details*, Washington, D.C., February 1999.

———, *45 Years of Global Reach and Power: The United States Air Force and National Security: 1947–1992, A Historical Perspective*, Washington, D.C., 1992.

———, *Toward the Future: Global Reach–Global Power: U.S. Air Force White Papers, 1989–1992*, Washington, D.C., January 1993.

U.S. Congress, Senate, "Force Structure Levels in the Bottom-Up Review," *Department of Defense Authorization for Appropriations for Fiscal Year 1995 and the Future Years Defense Program*, Washington, D.C.: Government Printing Office, March 9, 1994, pp. 687–753.

U.S. General Accounting Office, *Air Force Budget: Potential Reductions to Fiscal Year 1993 Air Force Procurement Budget*, Washington, D.C., GAO/NSIAD-92-331BR, September 1992.

———, *Bottom-Up Review: Analysis of DoD War Game to Test Key Assumptions*, Washington, D.C., GAO/NSIAD-96-170, June 1996.

———, *Bottom-Up Review: Analysis of Key DoD Assumptions*, Washington, D.C., GAO/NSIAD-95-56, January 1995.

———, *Defense Planning and Budgeting: Effect of Rapid Changes in National Security Environment*, Washington, D.C., GAO/NSIAD-91-56, February 1991.

———, *Defense Reform Initiative: Organization, Status, and Challenges*, Washington, D.C., GAO/NSIAD-99-87, April 1999.

———, *DoD Budget: Substantial Risks in Weapons Modernization Plans*, Washington, D.C.: GAO/T-NSIAD-99-20, October 8, 1998.

———, *Force Structure: Issues Involving the Base Force*, Washington, D.C., GAO/NSIAD-93-65, January 1993.

———, *Future Years Defense Program: DoD's 1998 Plan Has Substantial Risk in Execution*, Washington, D.C., GAO/NSIAD-98-26, October 1997.

———, *Future Years Defense Program: Risks in Operation and Maintenance and Procurement Programs*, Washington, D.C.: GAO-01-33, October 2000.

———, *Military Readiness: Data and Trends for January 1990 to March 1995*, Washington, D.C., GAO/NSIAD-96-111BR, March 1996.

———, *Military Readiness: Data and Trends for April 1995 to March 1996*, Washington, D.C., GAO/NSIAD-96-194, August 1996.

———, *Military Readiness: Updated Readiness Status of U.S. Air Transport Capability*, Washington, D.C., GAO-01-495R, March 16, 2001.

———, *National Security Issues*, Washington, D.C., GAO/OCG-93-9TR, December 1992.

———, *National Security: Perspectives on Worldwide Threats and Implications for U.S. Forces*, GAO/NSIAD-92-104, April 16, 1992.

———, *1992 Air Force Budget: Potential Reductions to Aircraft Procurement Programs*, Washington, D.C., GAO/NSIAD-91-285BR, September 1991.

———, *1993 Air Force Budget: Potential Reductions to Research, Development, Test, and Evaluation Programs*, GAO/NSIAD-92-319BR, September 1992.

———, *Quadrennial Defense Review: Opportunities to Improve the Next Review*, Washington, D.C., GAO/NSIAD-98-155, 1998.

———, *Strategic Mobility: Late Deliveries of Large, Medium-Speed Roll-on/Roll-off Ships*, Washington, D.C., GAO/NSIAD-97-150, June 1997.

Vick, Alan, et al., *Preparing the U.S. Air Force for Military Operations Other Than War*, Santa Monica: RAND, MR-842-AF, 1997.

Warner, Edward L. III, testimony before the House Armed Services Committee, February 2, 1994.

———, testimony before the Senate Armed Services Committee regarding structure levels in the Bottom-Up Review, March 9, 1994.

Watson, George M., and Robert White, end-of-tour interview with General Merrill A. McPeak, Air Force Chief of Staff, conducted at the Pentagon, November 28 and December 15 and 19, 1994.

White House, *National Security Strategy of the United States (NSS)*, Washington, D.C., August 1991,

———, "Remarks by the President to the Joint Chiefs of Staff and Commanders in Chief of the U.S. Armed Forces," September 15, 1998.

Wilson, George C., *This War Really Matters: Inside the Fight for Defense Dollars*, Washington, D.C.: Congressional Quarterly Press, 1999.

Wisner, Frank G., and David E. Jeremiah, "Toward a National Security Strategy for the 1990s," Washington, D.C.: Office of the Under Secretary of Defense for Policy, April 21, 1993.

Zakheim, Dov S., "Global Peacekeeping Burden Strains U.S. Capability," *Defense News*, April 6, 1998, p. 19.

———, "A New Name for Winning: Losing," *New York Times*, June 19, 1993, p. 21.

RECOMMENDED READING

Air Combat Command, Director of Combat Weapon Systems, *Ten Year Lookback: Standards and Performance, FY90-99*, February 4, 2000.

Aspin, Les, *An Approach to Sizing American Conventional Forces for the Post-Soviet Era: Four Illustrative Options*, House Armed Services Committee white paper, February 25, 1992.

———, confirmation hearings before the Senate Armed Services Committee, January 1993.

———, *Defense 1997 Alternatives*, briefing to the House Armed Services Committee, February 25, 1992.

———, *National Security in the 1990s: Defining a New Basis for U.S. Military Forces*, paper presented to the Atlantic Council of the United States, January 6, 1992.

———, "The Use and Usefulness of Military Forces in the Post–Cold War, Post-Soviet World," in Richard Haas, *Intervention*, Washington, D.C.: Carnegie Endowment, 1994.

Berner, Keith, *Defense Budget for FY1993: Data Summary*, Washington, D.C., Congressional Research Service Report 92-162F, March 18, 1992.

Biden, Joseph, "The Situation in Serbia," comments reported in the *Congressional Record*, Vol. 145, No. 4, S3047, March 22, 1999.

Blechman, Barry, and Steven Kaplan, *Force Without War*, Washington, D.C.: Brookings, 1978.

Bolton, John R., "Wrong Turn in Somalia," *Foreign Affairs*, January–February 1994.

Brown, Marjorie Ann, *United Nations Peacekeeping: Issues for Congress*, Washington, D.C.: Congressional Research Service Issue Brief 90103, updated December 20, 1996.

Bush, George H. W., "Address to the Nation on the Situation in Somalia," December 4, 1992.

———, "In Defense of Defense," August 1990 speech at the Aspen Institute, in Department of Defense, *Annual Report to the President and Congress*, Washington, D.C., 1991.

———, "Joint Statement with Prime Minister John Major of the United Kingdom on the Former Yugoslavia," December 20, 1992.

———, "Letter to Congressional Leaders Transmitting the Report on the Defense Management Review," July 10, 1989.

Clinton, Bill, "President Clinton's remarks at the U.S. Naval Academy graduation ceremony, May 25, 1994.

Clinton/Gore Campaign, "Clinton/Gore on National Security," undated fact sheet.

"Clinton Signs Defense Bill Despite Budget Increase," *1996 CQ Almanac*, Washington, D.C.: Congressional Quarterly Press, 1994, p. 452.

Congressional Budget Office, *The Costs of the Administration's Plan for the Air Force Through the Year 2010*, Washington, D.C., December 1991.

———, *Implications of Additional Reductions in Defense Spending*, Washington, D.C., October 1991.

———, *Making Peace While Staying Ready for War: The Challenges of U.S. Military Participation in Peace Operations*, Washington, D.C., December 1999.

Correll, John T., "The Base Force Meets Option C," *Air Force Magazine*, Vol. 74, No. 4, June 1992.

Daggett, Stephen, *Defense Appropriations Since the Beginning of the 104th Congress*, Washington, D.C.: Congressional Research Service, October 2000.

———, *Defense Budget for FY1997: Major Issues and Congressional Action*, updated October 10, 1996.

———, *Defense Spending: Does the Size of the Budget Fit the Size of the Force?* Washington, D.C.: Congressional Research Service Report 94-199F, February 28, 1994.

Davis, Paul K., Richard L. Kugler, and Richard J. Hillestad, *Strategic Issues and Options for the Quadrennial Defense Review*, Santa Monica: RAND, DB-201-OSD, 1997.

Defense Science Board, "Memorandum to Under Secretary of Defense (Acquisition), Subject: Report of Defense Science Board Task Force on Tactical Aircraft Bottom Up Review," July 14, 1993.

Department of Defense, *Mobility Requirements Study*, Washington, D.C., January 1992.

———, *Procurement Programs (P-1), Department of Defense Budget*, various years.

———, *Program Acquisition Costs by Weapon System*, Washington, D.C., various years.

Department of the Army, *A Statement on the Posture of the United States Army, Fiscal Year 2001*, Washington, D.C., 2000.

Department of the Navy, *FY 2001 Budget Estimates; Shipbuilding and Conversion Procurement History; Number of Ships*, Washington, D.C., February 2000.

Devroy, Ann, "Collapse of U.S. Collective Action May Force Second Look at Bosnia," *Washington Post*, October 8, 1993, p. A24.

Gellman, Barton, "Aspin Gently Criticizes Powell Report; Overlap of Military Functions Needs Another Look, Secretary Says," *Washington Post*, March 30, 1993, p. A6.

———, "Clinton's 1994 Defense Budget, Out Today, Meets Goals for Cuts," *Washington Post*, March 27, 1993, p. A9.

———, "Defense Budget 'Treading Water'; Most Issues in Achieving Long-Term Clinton Cuts Remain Undecided," *Washington Post*, March 28, 1993, p. A1.

———, "Defense Program Exceeds Budget Target, Aspin Says," *Washington Post*, September 15, 1993, p. A16.

———, "Pentagon May Seek $20 Billion More; Aspin Outlines Cost of Restructuring," *Washington Post*, August 13, 1993, p. A1.

———, "Pentagon Plan Would Cut Reserve; Politically Powerful 'Citizen Soldiers' Agree to Aspin Deal," *Washington Post*, December 11, 1993, p. A1.

———, "U.S. Reconsiders Putting GIs Under U.N.; Concern over Somalia and Bosnia Prompts Backlash in Congress," *Washington Post*, September 22, 1993, p. A1.

———, "Wider U.N. Police Role Supported; Foreigners Could Lead U.S. Troops," *Washington Post*, August 5, 1993, p. A1.

Gellman, Barton, and John Lancaster, "U.S. May Drop 2-War Capability; Aspin Envisions Smaller, High-Tech Military to 'Win-Hold-Win,'" *Washington Post*, June 17, 1993, p. A1.

Gordon, Michael, "Making the Easy Military Cuts," *New York Times*, March 28, 1993, p. 22.

———, "Pentagon Fights Budget Officials over $50 Billion," *New York Times*, December 10, 1993, p. A1.

Goshko, John M., "Clinton Seen Calming Hill on Peace Keeping; Caution in Committing U.S. Forces Said to Defuse Confrontation on Presidential Prerogatives," *Washington Post*, October 2, 1993, p. A16.

———, "U.S. Lists Stiff Conditions for Troop Role in U.N. Peace Keeping," *Washington Post*, September 24, 1993, p. A19.

HQ ACC/LGP, *Air Combat Command, Director of Logistics, Quality Performance Measures (L-QPM) Users Guide*, December 18, 1995.

Jeremiah, David E., statement of Admiral David E. Jeremiah, Vice Chairman of the Joint Chiefs of Staff, before the Committee on Armed Services, United States House of Representatives, March 12, 1991.

———, statement of David E. Jeremiah, Vice Chairman, Joint Chiefs of Staff, before the Senate Armed Services Committee, April 11, 1991.

Joint Chiefs of Staff, *National Military Strategy of the United States of America; A Strategy of Flexible and Selective Engagement*, Washington, D.C., February 1995.

Joint Staff, J-8 Division, "Force for 2000," unpublished briefing presented to Secretary Aspin, Washington, D.C., May 8, 1993.

Kaminski, Paul G., "International Partnerships Beyond 2000," address to ComDef '96 Conference, Omni Shoreham Hotel, Washington, D.C., April 1, 1996.

Kaufmann, William, *Planning Conventional Forces 1950–80*, Washington, D.C.: Brookings, 1982.

Krepinevich, Andrew, "How Much Bang, How Many Bucks," *Washington Post*, August 23, 1993, p. A17.

———, "National Defense Panel Report: First Shot in the Debate over Transforming the U.S. Military," *Center for the Strategic and Budgetary Assessments Backgrounder*, December 1, 1997.

Lancaster, John, "Aspin Opts for Winning 2 Wars— Not 1-1/2—At Once; Practical Effect of Notion Is Uncertain amid Huge Military Budget Cuts," Washington Post, June 25, 1993, p. A6.

———, "Aspin Says 5-Year Strategy Requires $50 Billion More," *Washington Post*, December 11, 1993, p. A12.

———, "Does Fighting Mire U.S. Forces? Retaliation Commits Troops to Risky New Phase of Peace Keeping," *Washington Post*, June 13, 1993, p. A32.

———, "Pentagon Issues Plan for Future; Review Proposes Modest Changes," *Washington Post*, September 2, 1993, p. A1.

———, "Pentagon Spending Gap Revised to $31 Billion," *Washington Post*, December 18, 1993, p. A8.

Larson, Eric V., *Casualties and Consensus: The Historical Role of Casualties in Domestic Support for U.S. Military Operations*, Santa Monica: RAND, MR-726-RC, 1996.

Larson, Eric V., and John E. Peters, *Preparing the U.S. Army for Homeland Security: Concepts, Issues, and Options*, Santa Monica: RAND, MR-1251-A, 2001.

Lippman, Thomas W., "African Crises Test Limited U.S. Commitment: Pressure Builds for More Direct American Intervention as Five Nations Suffer Strife," *Washington Post*, June 13, 1993, p. A33.

McPeak, Merrill, handwritten note from General Merrill A. McPeak to Directorate of Operations (XO) on JDAM, May 1, 1991.

Meyerson, Adam, "Calm After Desert Storm: Dick Cheney on Tax Cuts, Price Controls, and Our New Commander in Chief," *Policy Review*, No. 65, Summer 1993.

National Defense Panel, "Assessment of the May 1997 Quadrennial Defense Review, 1997.

———, "National Defense Panel Calls for National Security Transformation," press release, December 1, 1997.

———, National Defense Panel Seminar, Crystal City, VA, April 29, 1997.

———, *Transforming Defense*, Washington, D.C., December 1997.

Office of the Assistant Secretary of Defense, "Defense Department Funds Historic Increase in Housing Allowance," News Release No. 006-00, January 6, 2000.

———, DoD news briefing, October 1, 1998.

"Pentagon Examines Its Post–Cold War Role, *1993 CQ Almanac*, Washington, D.C.: Congressional Quarterly Press, 1994, p. 452.

Powell, Colin L., *Chairman of the Joint Chiefs of Staff Report on the Roles, Missions, and Functions of the Armed Forces of the United States*, Washington, D.C., January 1993.

Rhodes, John E., statement of Lieutenant General John E. Rhodes, "Defense Information Superiority and Information Assurance—Entering the 21st Century," hearings of the House Armed Services Committee, February 23, 1999.

Rice, Donald B., "Report of the Secretary of the Air Force," in DoD, *Annual Report to the President and Congress*, Washington, D.C., January 1993.

Ringgold, Timothy, *Strategy Turned Upside Down: The Bottom-Up Review and the Making of U.S. Defense Policy*, Washington, D.C.: Center for Strategic and International Studies, March 21, 1996.

Schmitt, Eric, "Pentagon Is Ready with a Plan for a Leaner, Versatile Military," *New York Times*, June 12, 1993, p. 1.

———, "Plan for 'New' Military Doesn't Meet Savings Goal," *New York Times*, September 15, 1993, p. A21.

Siegel, Adam, *U.S. Navy Crisis Response Activity, 1946–1989*, Alexandria, VA: Center for Naval Analyses, 1989.

———, *The Use of Naval Forces in the Post-War Era: U.S. Navy and U.S. Marine Corps Crisis Response Activity, 1946–1990*, Alexandria, VA: Center for Naval Analyses, 1991.

Smith, R. Jeffrey, and Julia Preston, "U.S. Plans Wider Role in U.N. Peace Keeping; Administration Drafting New Criteria," *Washington Post*, June 18, 1993, p. A1.

Taibl, Paul, "The $60 Billion Defense Modernization Goal: What, When, How Risky? *Business Executives for National Security Issue Brief*, March 1998.

U.S. Air Force, "Enhancing the Nation's Conventional Bomber Force: The United States Air Force Bomber Roadmap," in "Bomber 'Roadmap' and Related Bomber Programs and the Tri-Service Standoff Attack Missile (TSSAM)," hearing of the Committee on Armed Services, U.S. Senate, June 17, 1992.

U.S. Army Concepts Analysis Agency, *Force Employment Study (FES)*, Bethesda, MD, February 1991.

U.S. Congress, Senate, Senate Record Vote Analysis, 106th Congress, 2nd Session, June 7, 2000, 5:07 p.m., Page S-4649 Temp. Record, Session Vote No. 120, Defense Authorization/Base Closures, Subject: National Defense Authorization Act for fiscal year 2001, S. 2549, McCain/Levin amendment no. 3197.

U.S. Department of State, "The Clinton Administration's Policy on Reforming Multilateral Peace Operations," *U.S. Department of State Dispatch*, May 16, 1994, Vol. 5, No. 20, pp. 315–321.

U.S. General Accounting Office, *Air Force Bombers: Options to Retire or Restructure the Force Would Reduce Planned Spending*, Washington, D.C., GAO/NSIAD-96-192, September 1996.

———, *Air Force Organization: More Assessment Needed Before Implementing Force Projection Composite Wings*, Washington, D.C., GAO/NSIAD-93-44, May 1993.

———, *Cruise Missiles: Proven Capability Should Affect Aircraft and Force Structure Requirements*, GAO/NSIAD-95-116, April 20, 1995.

———, *Defense Acquisitions: Need to Revise Acquisition Strategy to Reduce Risks for Joint Air-to-Surface Standoff Missile*, Washington, D.C., GAO/NSIAD-00-75, April 2000.

———, *Defense Infrastructure: Costs Projected to Increase Between 1997 and 2001*, Washington, D.C., GAO/NSIAD-96-174, May 1996.

———, *DoD Budget: Evaluation of Defense Science Board Report on Funding Shortfalls*, GAO/NSIAD-94-139, April 20, 1994.

———, *DoD Budget: Future Years Defense Program Needs Details Based on Comprehensive Review*, Washington, D.C., GAO/NSIAD-93-250, August 1993.

———, *DoD Competitive Sourcing: Savings Are Occurring, but Actions Are Needed to Improve Accuracy of Savings Estimates*, Washington, D.C., GAO/NSIAD-00-107, August 2000.

———, *DoD Competitive Sourcing: Some Progress, but Continuing Challenges Remain in Meeting Program Goals*, Washington, D.C., GAO/NSIAD-00-106, August 2000.

———, *Military Airlift: Options Exist for Meeting Requirements While Acquiring Fewer C-17s*, Washington, D.C., GAO/NSIAD-97-38, February 1997.

———, *Military Operations and Capabilities Issue Area Plan: Fiscal Years 1995–97*, GAO/IAP-95-3, March 1995.

———, *Military Prepositioning: Army and Air Force Programs Need to Be Reassessed*, Washington, D.C., GAO/NSIAD-99-6, November 1998.

———, *1996 DoD Budget: Potential Reductions to Operation and Maintenance Program*, Washington, D.C., GAO/NSIAD-95-200BR, September 1995.

———, *1997 DoD Budget: Potential Reductions to Operation and Maintenance Program*, Washington, D.C., GAO/NSIAD-96-220, September 1996.

———, *Observations on the Future Years Defense Program*, Washington, D.C., April 1991.

———, *Operation Desert Storm: Evaluation of the Air Campaign*, Washington, D.C., GAO/NSIAD-97-134, June 1997.

———, *Quadrennial Defense Review: Some Personnel Cuts and Associated Savings May Not Be Achieved*, Washington, D.C., GAO/NSIAD-98-100, April 1998.

———, *Quadrennial Defense Review: Status of Efforts to Implement Personnel Reductions in the Army Materiel Command*, GAO/NSIAD-99-123, March 1999.

———, *Strategic Bombers: Adding Conventional Capabilities Will Be Complex, Time-Consuming, and Costly*, Washington, D.C., GAO/NSIAD-93-45, February 1993.

———, *Weapons Acquisition: Precision Guided Munitions in Inventory, Production, and Development*, Washington, D.C., GAO/NSIAD-95-95, June 1995.

White House, *A National Security Strategy of Engagement and Enlargement*, Washington, D.C., July 1994, February 1995, and February 1996.

———, "Review of National Defense Strategy," *National Security Review 12*, March 3, 1989, declassified on July 16, 1998.

———, Office of the Vice President, "From Red Tape to Results: Creating a Government That Works Better and Costs Less," *Report of the National Performance Review*, White House press release, September 7, 1993.

Williams, Daniel, and Ann Devroy, "Defining Clinton's Foreign Policy; Space of Speeches Will Seek to Kill Suspicions of U.S. Retreat," *Washington Post*, September 20, 1993, p. A16.

———, "U.S. Limits Peace-Keeping Role; Administration Delays Policy on Putting GIs Under U.N. Banner," *Washington Post*, November 25, 1993, p. A60.

Wisner, Frank G., memorandum to Secretary Aspin, February 23, 1993, reported in *Inside the Air Force*, Washington, D.C., March 12, 1993, p. 16.